GOLD

GOLD

BRIAN KETTELL

BALLINGER PUBLISHING COMPANY

Cambridge, Massachusetts
A Subsidiary of Harper & Row, Publishers, Inc.

Published in 1982 by
Ballinger Publishing Company
54 Church Street
PO Box 281
Harvard Square
Cambridge
Massachusetts 02138
USA

Library of Congress Cataloguing
in Publication Data

Kettell, Brian
  Gold
    Bibliography: p.
    Includes index
    I. Gold.   I. Title
  TN760.K47   1981   332.63   81-12708

  ISBN: 0 88410 857 0   AACR2

ISBN: 0 88410 857 0

Typeset and printed in Great Britain

Contents

Chapter 4   THE ROLE AND IMPORTANCE OF GOLD IN THE
              INTERNATIONAL MONETARY SYSTEM

Chapter 5   GOLD: THE MAJOR FACTORS AFFECTING
              ITS SUPPLY, DEMAND AND PRICE

List of Figures

List of Plates

(Plate section between pages 152 and 153.)

List of Tables

Acknowledgements

The author welcomes the opportunity to acknowledge the generous assistance provided on selected chapters by:
Robert Beale, Executive Director of Samuel Montagu and Co. Ltd.,
Keith M. Woodbridge, Managing Director of the International Monetary Market,
Valeurs White Weld, S.A.,
David Marsh and Kenneth Marston of the *Financial Times*,
Mocatta Metals Corporation,
Swiss Bank Corporation, Zurich,
Professor Arthur Laffer, University of Southern California,
George Rowe and Len Waxman, Economics Lecturers at City of London Polytechnic.

In compiling chapters five and seven of this book I have drawn heavily upon the statistics and conclusions contained in the 1980 and 1981 editions of the annual survey of the gold market, *Gold 1980* and *Gold 1981*, produced by Consolidated Gold Fields Limited (editor, David Potts).

Particular mention must also be made of the very generous assistance provided by Mark Wood, Gold Analyst of Grievson Grant and Co.

Finally Lise Jensen did a yeoman job of typing the entire manuscript.

Responsibility for any error remains with the author.

Chapter One

GOLD: IMPORTANCE, SCIENTIFIC PROPERTIES AND METHODS OF EXTRACTION

Introduction

The story of man's search for gold is as long as recorded time. No other metal has so captivated the human mind, and no other natural substance has been the source of so much bloodshed and violence. At the same time, gold has been instrumental, to a far greater degree than most other commodities, in providing real benefits to civilization.

Gold turned men into explorers. It lured the Egyptians on fragile boats and perilous expeditions around the African continent. It drove the Cretan and Phoenician captains to the very edges of the world of antiquity. When Christopher Columbus discovered America in 1492 the first thing he asked the natives was: 'Where is the gold?'.

Gold made men into inventors. While working gold the Egyptians developed brilliant skills in metallurgy. While mining gold, the Romans created new mining techniques. And when alchemists tried to produce gold from test tubes, they came up with gun-powder and porcelain and were ultimately responsible for the development of chemistry.

Gold turned men into slaveholders and murderers. In order to possess it, the Egyptians subjugated the Nubians, the Romans conquered Gaul and Spain, and the Spanish conquistadors annihilated entire peoples in Central and South America.

Gold created empires. High cultures like Egypt flourished because of their opulent gold production. The fact that the exploration of gold supplies involved advantages in addition to the gold produced was well recognised by Joseph Stalin. Having familiarised himself in great detail with the development of the United States, Stalin felt that the same principle could be applied to the outlying regions of the Soviet Union. He was interested in gold, but what appealed to him in addition was the fact that fortunes made during the American gold rushes not only built individuals but had developed whole regions. 'At the beginning, we will mine gold, then gradually change over to the mining and working of other minerals, such as coal and iron', Stalin wrote.

Lack of gold has forced empires into ruin. Rome's dominion collapsed

1

when its gold was lavishly wasted in all directions. Spain lost its importance when the source of gold dried up in the New World.

Much is known of placer mining for gold on the surface and hard rock mining which involved penetrating deep under the surface of the earth. What is not widely remembered is that the labour of gold recovery was backbreaking, endless and sometimes fatal. Only those who mined knew that the actual process of mining was for the young, that it was a lonely and uncertain process. The advantages the miner experienced were rough 'outdoorsy' and, to the miner, of dubious value. While one could discover the magnificence of an unspoiled landscape in the quest for gold, one in fact had to endure immense privation, numbing cold, dreadful food, isolation from friends and family, forced habitation in the most bitter of circumstances and one had to be reconciled to the company of rough men and unpremeditated violence. Mining did, however, bring into play new skills, transforming men into underground labourers. It produced a strain of men whose strength and skill remain unmatched, creating simultaneously legends and millionaires.

The appeal of gold is not linked purely to western cultures. This fact was well recognised when, in 1967, American planes dropped millions of gold coloured leaflets over North Vietnam. The leaflets were captioned 'Reward – Fifty Taels of Gold – Reward' and explained that anyone helping a downed American flier could obtain this reward.

But what is gold, from where does it come and to where does it go, and why is it the most coveted of metals by governments, investors and industrialists alike? It is in order to find the answers to these questions that this book has been written.

The lustre and fine colour of gold have given rise to most of the words which are used to denote it in different languages. The word 'gold' is probably connected with the Sanscrit word *jvalita* which is derived from the verb *jval*, to shine. An alternative explanation is that it stems from the old English word, *geola* meaning yellow. A geological definition of gold is that it is a yellow, malleable and durable element that occurs naturally in a pure state. The basic properties of gold were well summarised at the opening of the seventeenth century. As Francis Bacon put it at this time: 'Gold hath these natures – greatness of weight, closeness of parts, fixation, pliantness or softness, immunity from rust, colour or tincture of yellow.' Bacon's summary encapsulates some of the factors which have influenced man's desire to possess gold.

In the beginning, gold's appeal was aesthetic and mystical: the shining yellow beauty of gold came to symbolise the majesty of the gods, and its permanence reflected their immortality. Ancient myths and legends cast it as the child of Zeus, a metal with which to adorn temples and to offer an appeasement to the Gods. The ease with which the metal could be worked resulted in its use for decoration, and its rarity soon ensured its value as evidence of wealth. Gold became a symbol of power and riches, and kings and commoners alike yearned to possess it.

It is possible to list some 150 substances which at various times and in various parts of the world have been invested with some universality of value, almost equally divided between the animal, vegetable and the mineral. Gold's scarcity in relation to other minerals is not the only reason for an age-long hunger which has sometimes amounted to a passion. This scarcity, combined with a universal demand, is what gives gold a unique world-wide attraction.

Apart from its scarcity, this world-wide demand is based on the following characteristics possessed by gold, all of which have resulted in gold being accepted as the noblest of metals.

Physical and Mechanical Properties of Gold

Colour

The majority of metals, with the exception of gold and copper, are a greyish colour varying from bluish grey (lead) to white (silver). Gold is the only metal which has a yellow colour when in mass and in a state of purity. The derivation of the chemical symbol for gold, Au, from the Latin, *aurum*, meaning 'shining dawn' can be traced back to gold's colour. Impurities greatly modify this colour; small quantities of silver lower the tint, copper raises it. Silver and platinum impurities make gold white; traces of copper give the metal a reddish colour and iron produces varying shades of green. One of the rarest forms is black gold, which contains traces of bismuth. The beauty of gold has resulted in its widespread decorative uses.

Density

Density is the relationship of a volume of metal to an equal volume of water. Gold, which is 19.32 times as heavy as water, volume for volume, is said to have a density, or specific gravity, of 19.32. Much of the world's gold can be concentrated and recovered by simple gravity treatment, taking advantage of gold's high specific gravity. In ancient times a sheepskin was commonly used to trap particles of the metal from stream beds and it could have been this practice that gave rise to the legend of Jason and his search for the Golden Fleece. Gold is so dense in volume that the roughly 100,000 tonnes man has mined since 4000 B.C. could be contained in a cube measuring only 57 feet on each side.

Gold weights in troy ounces are often equated with weights in avoirdupois ounces or pounds. The two systems of measurement are based on the same unit of measure, the grain, which can be used to convert weight from one system to the other. The two systems of measurement are British but derive their names from the French. The troy system is used to weigh precious metals and gems and is named after Troyes, France, where it originated. The avoirdupois system, used to weigh everything but precious metals, gems and drugs, takes its name from the French phrase, *avoir du pois* meaning goods of weight.

The following table illustrates the troy system:

1 troy pound	=	12 ounces, 240 pennyweight, 5760 grains
1 troy ounce	=	20 pennyweight, 480 grains
1 pennyweight	=	24 grains, 0.05 ounces
1 grain	=	0.042 pennyweight, 0.002285 ounces

In metric terms a troy pound is 0.373 kilograms; an ounce is 31.103 grams and a pennyweight is 1.555 grams.

Taking the density of pure gold at 19.32, then: 1 c.c. of pure gold weighs 19.32 grammes, or 0.6205 oz. troy; 1 cubic inch weighs 316.25 grammes, or 10.168 oz. troy; 1 cubic foot weighs 546.485 kg., or 17569.9 oz. troy. The volume of 1 kilogramme of gold is 51.81 c.c. or 3.162 cubic inches. The volume of 100 oz. troy is 161.16 c.c. or 9.835 cubic inches. The volume of 1 ton avoirdupois is 1.86 cubic feet.

Malleability

Malleability is the property of metals which allows them to be hammered and beaten out into forms, such as thin sheets, without cracking. This malleability has long been recognised. The historian Pliny observed: 'Nor is any other material more malleable or able to be divided into more portions, seeing that an ounce of gold can be beaten out into 750 or more leaves four inches square'. Leaves of not more than 1/300,000 of an inch in thickness can be obtained by beating gold. Even in ancient Egypt a skilful gold beater could hammer gold bullion to such a fine consistency that it would have taken 250,000 sheets to produce a layer one inch thick. Nowadays a single ounce can be beaten into a thin translucent sheet of about 10 square metres (108 square feet) in area and only some 120 millionths of a millimetre in thickness. Because it is malleable and does not tarnish, gold is used by dentists for crowns, inlays, bridges and partial dentures. These same properties also led the Chinese to use gold for acupuncture needles.

Ductility

Ductility is the property of metals which enables them to be given a considerable amount of mechanical deformation (especially stretching) without cracking. Gold is so ductile that one single ounce can be drawn into 50 miles of fine wire. If gold thread is required for gold lace or gold tissue (from which 'tissue paper', originally designed for placing between folds of gold-woven fabric, derives its name), an ounce of gold very thinly plated in leaf form, upon a silver or copper wire base will suffice for over 1,000 miles. In fact gold is so ductile that it can be made into wire so fine that a gold thread drawn from one ton of gold would stretch from the earth to the moon and back. The presence of traces of other metals, especially lead, reduces the ductility and malleability of gold. Malleability and ductility are possessed by gold at all temperatures to a far higher degree than by any other metal.

Indestructibility

Unlike silver, gold does not tarnish and is not corroded by acid – except by a mixture of nitric and hydrochloric acid. Gold is so firmly enthroned as 'king of the metals' that this mixture of acids by which it can be dissolved is known as *aqua regia*, or royal water. Gold is immune to the effects of weather, water and oxygen. It neither tarnishes, rusts nor corrodes.

The finest achievements of the goldsmith's art are to be found in ancient Egypt and in Mycenae, and have been unearthed by archaeologists in perfect condition. 'Gold is the Child of Zeus,' wrote Pindar 'neither moth nor rust devoureth it.' When the archaeologist Howard Carter discovered the tomb of Tutankhamen in Egypt in 1922, the golden treasures of the boy king who had reigned from 1361–1352 B.C. were perfectly preserved. The king's body was encased in a coffin of solid gold nearly 3 millimetres thick, weighing 242 pounds; the mummy had a portrait mask, the hands and toes were sheathed in gold. Among the rich assortment of possessions in the tomb was a sheath of solid gold, embossed with animal scenes. There was a golden throne with a back panel of exquisite workmanship showing the young king being anointed by his queen. These treasures, now in Cairo Museum, look almost as perfect as when they were first created three thousand years ago.

Gold is, however, attacked when it becomes the positive electrode of an electrolytic cell in a hydrochloric acid solution. This fact is the basis for an electrolytic refining process.

The atomic number of an element serves two functions. It shows the chemical properties an element must have due to its relationship to other elements in its group or period. In order to make jewellery, gold is often mixed with a little zinc, and rather more silver and copper, the two metals which gold most closely resembles in atomic structure (it shares their subgroup in the periodic table of elements). Secondly, it shows how that element may have been built up in the process of evolution before and after another element. The simplest atom, that of hydrogen with an atomic number of 1 and an atomic weight of 1.008, has a nucleus of one proton and one electron orbiting it. Gold, with an atomic number of 79 and an atomic weight of 196.97, has a nucleus of 79 protons and 118 neutrons. Around the nucleus are 79 electrons. It is the arrangement of these electrons around the nucleus which makes gold an almost indestructible shining metal.

Electric and Thermal Properties

The resistance to tarnishing and corrosion ensures constancy of surface properties, such as contact resistance over extended periods. Important among gold's properties are its electrical properties of resistivity (or conductivity), contact resistance, temperature coefficients or resistance and high thermal e.m.f. against copper. Metals are conductors of heat but they conduct it in varying degrees. Silver is the best and, using 100 as its conductivity, copper is the second best at 73.6 and gold the third best at 53.2.

Indeed gold is such an excellent conductor of electricity that a microscopic circuit of liquid gold 'printed' on a strip of plastic can replace miles of wiring in a computer. Moreover, gold's high thermal conductivity ensures rapid dissipation of heat when gold is used for contacts or in situations where constriction resistance arises.

Space exploration has provided exciting new uses for gold. In the conditions outside the Earth's atmosphere conventional lubricants can deteriorate rapidly, so hard gold alloys are used to coat vehicle bearings. The umbilical cord that tethered astronaut Edward White, the first American to walk in space, to his Gemini spacecraft was gold-plated to reflect thermal radiation. When men landed on the moon, gold foil shrouded the miniature television camera and other instruments on the moon buggies to protect them from the fierce rays of the sun in space. Gold's malleability is again important here, permitting fine layers of gold less than 0.005 millimetres (0.00012 inches) thick to be applied.

The fusion point of a metal is the temperature at which it melts and becomes a liquid. This varies greatly for different metals, with the melting point of gold being relatively low at 1064 degrees Centigrade (1947.2 degrees Fahrenheit). The melting point of gold constitutes a fixed point for pyrometry (the science of measuring high temperatures).

Where can Gold be Found?

Gold is widely dispersed throughout the Earth's crust. Naturally occurring metallic, or native, gold usually contains variable amounts of silver, copper, platinum, palladium, or certain other elements mixed with it. Gold exists in association with most copper and lead deposits, and although the quantity of gold present is often extremely small, it is readily recoverable as a by-product in the refining of those base metals. Gold is also found in seawater. Although the total amount in the ocean is estimated at billions of tons, the concentration of it is less than 6 parts per 1,000,000,000,000 parts of seawater, making it not economically feasible to mine the sea.

Some appreciation of the problems and costs of extracting gold from the earth can be perceived from the following list of ingredients needed to produce one fine ounce of gold: 3.7 metric tons of ore, 38 man hours, 6,400 litres of water, sufficient electric power to run an average household for ten days, between 8 and 17 cubic metres of compressed air and varying quantities of stores and chemicals including cyanide, zinc, acids, lime and borax.

The two main types of native gold, i.e. gold in its free elemental state, are 'alluvial' and 'reef' (sometimes referred to as 'lode'). The circumstances under which these occur are described directly below.

Alluvial Deposits

Alluvial deposits, or 'placers', are formed by mechanical process and from the weathering and disintegration of mineral-bearing rocks and veins, followed by the transportation and concentration of the freed mineral by the action of running water, the most efficient means of separating light from heavy minerals. Metallic and non-metallic ores which are highly resistant to weathering become separated from their more easily weathered parent rock. The main agents which bring about the disintegration are variations in temperature, the expansion or freezing of water, erosion due to the movement of water, and the sandblasting action of the wind. The best conditions for the concentration of gold in auriferous (gold bearing) gravels are when the river gradient is moderate, under balanced conditions of erosion and deposition.

Placer deposits may be separated into five main types:

(i) Eluvial (or residual) deposits include the so-called hillside occurrences, which are formed by the weathering of outcrops, and are usually located, *in situ*, on the gentle slopes of valleys. The gold-encasing material is most effectively broken down by persistent and powerful geological conditions which effect the mechanical breaking down of the rock and the chemical decay of the minerals. The surface of a goldbearing ore body is enriched during the process of rock disintegration, because some of the softer and more soluble parts of the rock are worn down by erosion. After gold is released from its bedrock encasement by rock decay and weathering it begins to creep down the hillside and may be washed down rivulets and gullys and into the stream beds. On its way down the hillside the gold is sometimes concentrated in sufficient value to warrant mining.

(ii) River and stream placers include creek and river bed deposits which are formed in, or bordering, rivers and streams and are generally found where the velocity of water current has slackened, thus decreasing the carrying capacity of the water. The alluvial load consists of sand, coarse gravel (often exclusively of quartz pebbles) and freed minerals which have resisted erosion and weathering. The native gold and heavy black (magnetic) sands are deposited in the valleys and draining channels, according to the specific gravity and grain size.

Because of its insolubility, the gold remains unaltered and unaffected by chemical change. Stream placers were the source of most of the placer gold mined in California.

(iii) River terrace, or bench deposits, occur on the flanks of valleys and are remnants of old river deposits, formed by the river cutting itself into a deeper channel in the bedrock.

(iv) Beach or marine placers occur along coastal strips in many countries, having been formed by the sorting action of the waves. This tends to concentrate beach material resulting from the erosion of rocks together with any heavy minerals which may have resisted weathering.

(v) Deep leads, or buried placers, are ancient deposits which have been buried under overburden, varying in thickness from 100 to 1,000 feet and more. There are several ways that a placer deposit could be preserved. Since streams are constantly changing their position, fragments of their deposits are left isolated; for example, the beaches and terraces that are left at different intervals when a stream is cutting a deeper channel. Those deposits that are left will eventually be eroded unless something protects them. Burial is the most effective way a placer may be preserved. Mostly when the name 'buried channel' is given to a placer it is one where a stream has been covered by lavas, ash falls, landslides, glacial material or marine sediments.

The character of alluvial gold is variable but closely related to the parent ore body. Placer gold occurs in flat scales and flakes (paillettes), in rounded particles (pepites) and as irregularly-shaped grains, rolled masses and nuggets bearing evidence of much attrition. Some of the largest masses of gold ever found have been located in placer deposits. Alluvial gold has generally a higher degree of fineness than vein gold and contains a correspondingly smaller percentage of silver.

Reef Gold

Reef gold is a native metal embedded within arteries of ore or quartz. It is the parent ore body of alluvial deposits and has to be obtained by mining. Appendix I (*see p. 13-17*) gives a detailed account of how gold is mined from a typical South African gold mine.

The Mining of Gold

There is no single mining method applicable to all gold deposits. Gold is recovered by placer mining of alluvial deposits, and by lode or vein mining. It is also recovered as a by-product of base-metal mining. Placer mining, the oldest method, entails exploiting the high density (or heaviness) of gold to separate it from the much lighter siliceous material with which it is found. The alluvial deposits mined by placer methods are the gold-bearing sands and gravel that have been deposited by rapidly moving streams and rivers at places where they widen or, for some other reason, lose speed. As the current slows, the sediment being carried downstream settles to the bottom.

Placer Mining

Although the basic principles of placer mining have not altered since early times, the methods have improved considerably, chiefly in mechanical processes. In the great American gold strikes in California, Colorado and Alaska, discussed in Chapter 2, placer mines were almost exclusively the source of gold and the panning method was the technique utilised by the individual miners. The miner used a pan or a batea (a pan or basin with radial corrugations), in which he placed a few handfuls of dirt and a large amount of water. Swilling the pan washed the siliceous material over the side and left the denser material in the centre of the pan. After many washings, only gold and the other heavy minerals were left. At this point, if the gold particles were large enough, it was comparatively easy to separate them from other materials.

With a specific gravity of 19.32, six or seven times that of quartz, the gold works its way to the bedrock or to some point where it can go no further. Once it is trapped on bedrock the stream has great difficulty picking it up again. How rich the deposit is will depend on how complete is the loss of transporting power, as well as the ability of the bedrock to hold the deposited gold, and the general relationship of the gold sources to the stream. A smooth hard bedrock is a very poor one to develop a good placer deposit. The bedrocks that are highly decomposed and possess cracks and crevices are good sources for gold and those of a clayey or schistose nature are rated excellent in their ability to trap the gold.

The cradle, or rocker, was an improvement over the pan and the batea. The cradle, named from a child's cradle, which it resembled, could process larger quantities of ore. Gravel was shovelled on to a perforated iron plate, and water was poured in. The finer material dropped on to the apron which distributed it across riffles, pieces of wood or iron perpendicularly fixed to the bottom and sides of the cradle. The entire apparatus was rocked, and as the material moved through the rocker, the gold was caught by the riffles. When enough gold was accumulated, the riffles were cleaned.

Hydraulic Mining

In California, thick beds of gravel on the hillsides were worked by hydraulic mining. Powerful jets of water at pressures of hundreds of pounds per square inch were passed through giant swivel-mounted nozzles to break down the gravel banks and wash the material through lines of sluices. Although great volumes of water and many miles of pipes and flumes (artificial channels used for conveying water) were required, the cost of treatment was only a few cents per cubic yard, which made it possible to work even poor ground at a profit. The millions of tons of tailings (discarded residue) that were washed into the Yuba and Feather rivers, however, had such an adverse effect on farming downstream that an injunction was obtained against hydraulic miners in 1880, and the work thereafter was strictly limited.

Dredging

In the early 1900s, dredging became the most important type of placer mining and has become probably the single most prevalent technique. The dredge generally used the world over is similar to that employed to deepen harbours and rivers. Invented in New Zealand, the gold-mining dredge achieved its greatest popularity on the rivers there and in California. Dredging is the major technique used today in the USSR.

Paddock dredging, a later development in the western USA, makes it possible to work placer deposits even if they are not in or near a riverbed. The paddock dredge floats in a pond that is continuously extended by the digging equipment at one end of the dredge while simultaneously being filled in by the waste or tailings at the other end. In this way, the dredge moves across country, taking its pond or reservoir with it. By piling more gravel around the reservoir and increasing the water level, the dredge can be made to work its way uphill. In 1910 there were 72 operating dredges in California. After World War II, however, only a few remained in operation.

The largest gold dredge in the world today operates on a river near Lake Baikal in the USSR and measures 236 metres (775 feet) overall, towering more than 40 metres (130 feet) above the water. It is believed to be capable of excavating about 2000 tonnes of gold-bearing gravel per hour from a depth of 50 metres (165 feet).

Vein or Reef Mining

Many of the methods used in the underground mining and exploration of gold lode or vein deposits are similar to the shaft- and pit-mining methods used for other metals. Gold-bearing ore tends to be extremely hard and abrasive, formed from ancient pebbles and sand that have been fused together by extremes of heat and pressure. In recent years great effort has been made to develop rock cutting machines, thus allowing the reef and waste to be extracted without the use of explosives, but an economic system is not yet available. Rock breaking underground is still performed almost exclusively by the traditional approach, using pneumatic drills and tungsten carbide tipped steel to drill holes which are then charged with explosives for blasting. Great tonnages of gold ore are treated throughout the world since most gold-mine ores contain an extremely low percentage of gold. For example, in one area three tons of ore are processed for every ounce of gold obtained. As described in Appendix I (*see pages 13-17*) the broken rock is loaded and transported by locomotive to the shaft system and finally hoisted up to the surface for dumping or processing.

Some 98 per cent of South African gold comes from underground operations. The South African gold miners have to overcome enormous physical problems and hazards in their search for gold deeper and deeper underground. In one mine, in the Transvaal, workings now extend to a depth of 14,000 feet (2.6 miles) and miners have to descend in stages as no lift cable can support its own weight to such depth. At this depth the rock is so hot as to burn the skin. Indeed, at a depth of 2,200 metres the temperature of the natural rock (virgin rock temperature) is 45 degrees Centigrade and rock temperature rises some 10 degrees Centigrade for every additional 1000 metres depth. Consequently refrigeration systems are widespread in the industry to maintain air temperatures at tolerable levels. The environment also tends to become humid as water is used to allay the dangerous siliceous dust generated by mining.

Strata control – ensuring that underground excavations remain open, allowing the extraction of ore in a safe and efficient manner – presents difficult problems in the deep South African mines. The phenomenon of the rock burst, which accounts for almost 80 per cent of the overall accident fatality rate in gold mines, disrupting production, causing equipment losses and having a detrimental effect on labour recruitment, is the most serious threat. A rock burst must be distinguished from a rock fall. A rock burst has been defined as 'a seismic event radiating sufficiently intense shock waves to cause visible damage to an excavation'. A rock fall is the simple falling or scaling of rock chunks from the roof or sides of an excavation.

Recovery and Refining of Gold

In the late nineteenth century amalgamation with mercury was the main method used to separate gold from its ore. Under this method the crushed rock was passed with water over mercury treated copperplates which trapped the gold particles. The plates were then cleaned off from time to time and the gold separated by distilling off the mercury. By the 1890s, however, problems were being experienced in the South African gold mines as workings deepened and the sulphide content of the ore increased. Near the surface the reefs were rich and oxidised and recovery rates as low as 50 per cent were profitable, but the presence of iron sulphides made the amalgamation process less effective.

Cyanide Process

The cyanide process, introduced in South Africa in 1890, effected a vast improvement over amalgamation and other earlier methods, and has been extensively used ever since. In this process, the gold in finely ground ore is dissolved by treating it with a very dilute solution of sodium chloride or less expensive calcium cyanide, plus lime and oxygen from the air. The mixture is held for some hours in large tanks equipped with agitators. The chemical reaction yields a water solution of gold cyanide and sodium cyanoaurite. This solution of gold is treated to remove oxygen, then clarified and mixed with zinc dust to precipitate the gold and the other metals, such as silver and copper, that were dissolved by the cyanide. The precipitate is then treated with dilute sulphuric acid to dissolve residual zinc and most of the copper. The residue is washed, dried, and melted with fluxes (materials used to promote fusion of the gold and silver and to dissolve the remaining copper). The operation may be repeated to flux off more base metal. The remaining gold and silver alloy, called *doré*, is then cast into moulds for assay.

The cyanide process may constitute the whole recovery process or may follow amalgamation. At some mines in the USA amalgamation is used to recover about 60 per cent of the gold, after which cyanidation achieves an overall recovery of 95 to 96 per cent. The cyanide process has been used to treat the tailings from early operations and fortunes have been made in reworking old dumps. The process is employed almost exclusively in recovering gold from the South African mines.

The most important and exciting recent technique for the recovery of gold is what is known as the 'carbon-in-pulp' process. This is a relatively simple process. Gold is dissolved in cyanides in the usual way but, instead of filtering off the liquid and precipitating gold from the solution by using zinc, carbon is added in the form of broken-up charred coconut shells. The gold adheres to the carbon and the gold and carbon are then separated from the pulp by screening. Rates of extraction as high as 99 per cent are claimed to have been achieved.

The properties of gold in ores from the standpoint of recovery are its extremely high specific gravity (15.5 to 19.32 depending upon the amount of alloying metal admixed), the fact that mercury wets it readily in the presence of water (amalgamation), its solubility in dilute aqueous solutions of alkaline cyanides to form relatively stable compounds, and its response, particularly as naturally alloyed to flotation collectors.

Recovery from Scrap

Considerable amounts of gold are recovered from scrap. The composition of scrap varies widely and dictates the process to be used. The scrap is largely produced in the manufacture of gold jewellery and electrical contacts, and as electroplating residues. Melting the scrap under oxidising conditions will volatilise (evaporate) some of the zinc and remove some of the iron in the slag. The metal may then be formed into pellets and if the gold content is about 20 per cent it can be treated with nitric acid to remove the base metals. If the pellets are low in lead, treatment with hot sulphuric acid may suffice. Retreatment will be required, however, to yield gold of acceptable purity if the sulphuric acid process is used. These methods are appropriate for scrap that is free of platinum metals. Lower-grade scrap may be melted and cast into anodes that are electrolysed in a sulphate solution. The gold will remain in the anode mud and can be recovered by first converting it to gold chloride and then precipitating the gold by adding ferrous sulphate.

Refining

In the Wohlwill Process, *doré* gold from the usual mining and smelting operations is electrolysed in a chloride solution using direct plus alternative currents. The gold in the *doré* anode dissolves and is deposited on the cathode as very pure gold. The silver is converted into chloride of which some clings to the anode and must be removed from time to time. This material also contains considerable gold and must be reworked. Any platinum or palladium in the anode will dissolve and is recovered by treating the electrolyte. This extensively used process yields gold at least 99.95 per cent pure.

Because the Wohlwill Process is rather slow, and for other reasons, it has largely become replaced by the Miller Process, in which chlorine is bubbled through the molten *doré*, converting the base metals into chlorides, some of which volatilise (i.e. evaporate). The silver also is converted to silver chloride, which is molten and can be poured off and recovered. After this treatment, the remaining gold usually has a purity of 99.5 per cent or above.

APPENDIX I*

HOW GOLD IS PRODUCED ON A SOUTH AFRICAN MINE

Gold is located in thin layers or sheets of rock called 'reefs' which are inclined at an angle from the horizontal plane and lie thousands of feet below the earth's surface. Figure 1.1 shows how the gold ore is mined and the treatment which is necessary to extract the gold from the rock in which it occurs, usually as minute particles and, more rarely, as visible gold flecks. After the headgear or framework which carries the winding ropes, pulleys and other mining equipment have been installed on the surface, a shaft is sunk beneath this structure which eventually may go over a mile deep to reach the areas of gold bearing rock (1). From this shaft, at various depths, passages radiate, like nerves from a central core, towards the reefs (2). These passages are excavated ('developed' in mining jargon) by drilling a pattern of holes at various angles into the rock which are then filled with explosives and timed to go off at different intervals (2A). The process of developing passages like these through waste or barren rock is called 'waste development'. This is necessary either to reach the roof or for other purposes such as ventilation or the transportation of rock and equipment through the mine. When the passages reach the gold bearing rock they branch out and are developed on the plane of the reef, which

*(Reproduced by kind permission of Consolidated Gold Fields Limited)

might vary in thickness from only a few inches to five or six feet but can cover many square miles in area. This is called 'reef development'. The gold bearing rock is then removed rather like withdrawing a slice of meat from a sandwich. This process of mining out the reef is called 'stoping' (3). While stoping continues, more reef and waste passages are developed. These form a network of tunnels which run for miles, enabling new areas to be mined as old areas become depleted of ore.

Waste rock from waste development and ore from stoping and reef development (which inevitably contains some waste) are transported to tipping stations separately and dropped down the appropriate rock or waste chute to a central collecting point before being hoisted up the shaft and placed into separate storage bins on the surface (4). From here trains deliver the waste rock to a dump (5), while the ore is taken to the mill and fed into a primary jaw crusher (6). This crusher has a similar action to a nut cracker, breaking down the larger pieces of rock to just under five inches in diameter. The lumps of rock then pass along a conveyor belt and any pieces not containing gold bearing reef are removed by persons trained as sorters (7). Investigations are now in progress to replace this manual sorting by mechanical methods. The pieces of ore are then reduced to a size of under ¾ inch in diameter by secondary gyratory crushers (8). In these, a truncated cone gyrates inside a stationary cone, crushing the ore against the inner walls. After passing through the second gyratory crushers, water is added to the ore and the mixture is fed into a number of ball mills (9). These are horizontal revolving steel cylinders partially filled with steel balls which grind all the material to a coarse pulp. The pulp then goes into a battery of tube mills loaded with selected pieces of reef rock where, by much the same process as the ball mill, it is crushed further until it becomes a fine slime (10). This is allowed to settle in tanks (11) and any surplus water is drained off and re-used in the mills. When settled, the thickened portion is drawn off from the bottom and fed into agitator tanks where a cyanide solution is added to dissolve the gold (12). The process of dissolution is helped by adding oxygen in the form of compressed air which is blown up through the centre of the tanks; another useful function of the air is to agitate the solution ensuring thorough mixing with the cyanide.

The waste solids are separated from the gold bearing solution by a rotary filtration process (13). This involves the slime from the agitator tanks passing first into a bath where there is a partially immersed rotating drum covered with a fine filter cloth. The gold bearing solution is sucked through the cloth by pipes which radiate from the centre of the drum. The solids which accumulate on the outside of the cloth are scraped off the drum just before re-entering the bath, mixed with water, and pumped on to the slimes dump as waste (13A). The gold bearing solution is then sucked through a candle filter and the suspended waste solids collected on the candles, which are PVC tubes punctured with small holes and covered with closely bound nylon line (14). It is now necessary to precipitate the gold from the solution which must first have the oxygen removed from it. The air is removed by passing the solution through a vacuum chamber. The precipitation process is completed by adding zinc dust to the de-aerated solution (15). The precipitated gold-zinc solution is poured into tanks containing filter leaves, which are open frames covered with canvas (16). A vacuum pump draws the solution through the canvas leaving a gold-zinc sludge caught on the cloth, while the barren liquid is returned to the agitator tanks.

The latest method used on 50 per cent of the mines, is for the precipitated gold-zinc solution to be drawn from the vacuum chamber (15) directly through the precipitation filter (17) whereby the solids are caught up in the candles, in the same way as illustrated in diagram 14. This eliminates stage 16 shown on the diagram. The precipitation candle filters are backwashed to remove the gold sludge from the candles. This is then passed through a filter press (18) which dewaters the sludge. Here the sludge is caught between a series of filter leaves while the solution passes through. When the gold sludge has built up and filled the spaces between the filter leaves, the leaves are pressed together and the water expelled. The press is then emptied by opening it and scraping the gold sludge off the leaves into a tray below. An alternative method of dewatering is to use a rotary filter (as shown in diagram 13). The resultant gold mixture goes to the smelt house where it is first roasted in a calcining furnace which removes moisture and oxidises impurities, the zinc is withdrawn as a vapour (19). After this the calcined slime is mixed with certain 'fluxes' which are materials such as borax and sand which help separate the gold from the impurities forming a waste scum or slag. The fluxed gold slime is then smelted in a submerged arc furnace for about two hours. The furnace is then tilted and the molten gold and slag poured into bar moulds arranged in a cascade (20). The gold ingots from the moulds weigh approximately 1,000 oz. and contain 90 per cent gold and 10 per cent silver. The gold bars are then sent to the Rand Gold Refinery near Johannesburg where the silver is extracted and the gold refined to a minimum of 99.5 per cent purity.

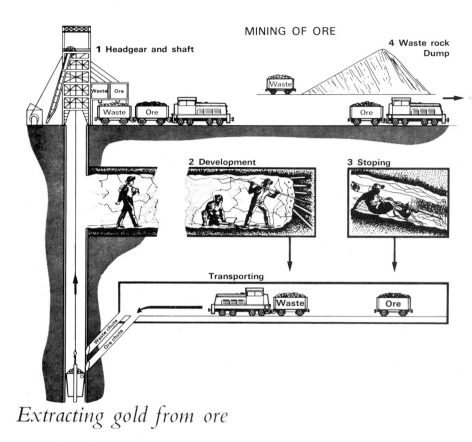

Extracting gold from ore

Figure 1.1. How gold is produced on a South African mine.

METALLURGICAL PROCESS

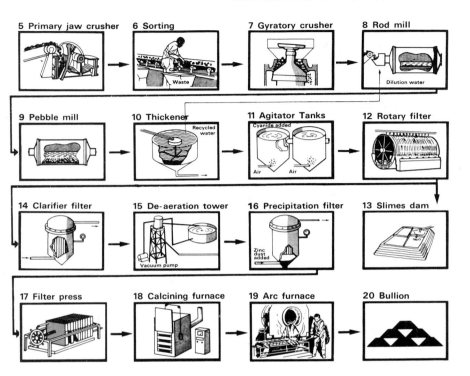

Chapter Two

EARLY HISTORY,
THE GREAT GOLD RUSHES, AND
THE DEVELOPMENT OF COINAGE

Introduction

The origins of money are lost in antiquity. The ability of money to free individuals from the cumbersome necessity of barter must have led to its early use as soon as some generally accepted commodity appeared. Man has used just about everything one can imagine as money ever since he needed a medium of exchange to make trading easier. Among the items have been beads, pretty seashells, feathers, large stones, tobacco, and cattle, – the list is a long one. Generally the 'money' was something valued for its beauty or its usefulness to nearly everyone, but also it was usually something that man could not quickly produce in great quantities, thus inflating the money supply. The tobacco standard in seventeenth century Virginia ceased being effective when a large part of the population owned their own mints, i.e. cultivated their fields of tobacco. Modern sea dredging put an end to the cowry shell standard of East African peoples.

Precious metals asserted their ascendancy as a medium of exchange and a store of value very early on in man's development and soon came to circulate as money. They were in heavy and permanent demand for ornament and decoration, and they were in continuous supply (since they did not easily wear out). Thus they tended to have a high and stable price. They were easily recognised and generally known to be commodities which, because of their stable price, would be accepted by most people. They were also divisible into extremely small units (gold, to a single grain). As mentioned in Chapter 1 the 'grain' is actually still used in the Avoirdupois system and in the Troy system. One grain is equivalent to 0.0648 grams, 24 grains make one pennyweight (dwt), and one pound troy consists of 5760 grains.

The Early History of Gold

Gold is the aristocratic, or royal metal without equal, as is convincingly shown in the records of history and archaeology. There are numerous Biblical references to the existence and to the scale of ancient wealth of gold in the Near East. In fact gold is the first element mentioned in the Bible. Quoting from Genesis (2:2–10) we read:

'A river flowed out of Eden to water the garden and there it divided and became four rivers. The name of the first is Pishon; it is the one which flows around the whole land of Hav-i-lah, where there is gold; and the gold of that land is good'.

In the earlier Biblical period, gold was used extensively as a store of value, but its main use was decorative rather than monetary. Kings V records that an emissary sent by the King of Syria to the King of Israel 'took with him ten talents of silver and six thousand pieces of gold'. Kings IX also records that Hiram, King of Tyre, supplied Solomon with gold for the decoration of the Temple of Jerusalem on two occasions, to the amount of a hundred and twenty and four hundred and twenty talents respectively. The second amount was said to have come from the mysterious gold producing land of Ophir, the location of which is still unknown. It seems that the Temple as well as being used as a treasury, was also used as a bank, at any rate to the extent of receiving money for safe keeping. The lavish use of gold for decorative purposes in the Temple served partly the object of accumulating a monetary reserve for any future emergency. When Hezekiah had to pay a tribute of three hundred talents of silver and thirty talents of gold to the King of Assyria (around 700 B.C.) he 'cut the gold from the doors of the Temple of the Lord, and from the pillars'.

The Queen of Sheba is also recorded as having given Solomon 'one hundred and twenty talents of gold'. Solomon, we are told in the book of Kings, was fortunate enough to have an annual income of 'six hundred three score and six talents of gold'. It has been estimated that the minimum equivalent weight of a talent was about sixty pounds (twenty-seven kilograms); little wonder, then, that Solomon 'exceeded all the kings of the earth for riches'.

Although there is evidence of some exaggeration in the actual physical quantities transacted, the many Biblical references to gold leave no doubt as to its widespread usage and acceptability. History further records Pythius, the subject of Xerxes of Persia, with his treasure of some 7,000 lbs. of gold; upwards to 500,000 lbs. of gold captured by Emperor Trajan from the Dacians, and the treasure of 320,000 lbs. of gold left by the Byzantine Emperor Anastasius at his death in 518 A.D.

The two oldest civilisations known to us in any detail are those of Egypt and Mesopotamia which were 'gold based' and, by the very nature of things, were founded on chance discoveries of the metal. Both were valley cultures, relying on a great river system for the concentration of the large alluvial gold deposits. History records that the earliest source of gold came from the first culture known, Sumeria (modern day Iran and Iraq). In the fifth millenium B.C. the Sumerians controlled the area between the

Euphrates and the Tigris and, as well as developing cuneiform writing and discovering the potter's wheel, also worked gold. This gold was treasured not for its monetary value but for the divinity it bestowed on the possessor. The sacred instruments of temples were made of gold and so were the insignia of earthly deities, the kings and the priestesses. Leonard Woolley excavated the royal tombs in the Sumerian capital of Ur and discovered much evidence of Sumerian goldsmiths' skills in the working of gold.

Around the same time that the Sumerians were in Mesopotamia, Egypt, the richest gold land in early antiquity, was experiencing an economic boom. In Egypt, the gold poured out of the earth and the rivers. There was gold on the shores and in the bed of the Nile.

The Egyptians developed great skill in the mining of gold and also invented the arts of gold smelting and refining. So great, however, was the Pharoah's demand for gold that Egyptian soil would not supply enough. The rulers sent their gold minded geologists to Arabia, Cyprus and the Far West, and it is thought that on one of their voyages the Egyptians circumnavigated Africa for the first time. The divine Kings, however, had their most significant gold suppliers nearby: this was the land that began at the southern border of Egypt and was soon named, following its occupation, after the Egyptian word for gold (*nub*), Nubia. The gold history of Egypt lasted for 4,000 years ending with the beginning of the Christian era.

Egypt was the richest gold producing area of the ancient world until the systematic Roman exploitation of Spanish gold deposits started at the beginning of the Christian era. The Graeco-Roman period witnessed the use of gold as coinage for the first time rather than its previous role as ornamental and decorative. The political unification of Greece was effected by Philip of Macedonia in 336 B.C. The subsequent subjugation of the Persian Empire, previously the dominant world power, by Philip's son, Alexander the Great, resulted in Greek control over a nation which had acquired a large proportion of the world's gold supply. As a result of this victory, the centre of gravity of gold, which until then had been contained in an area bounded in the south by Egypt, on the West by the coast of Asia Minor, and on the North by the Black Sea, was able to move slowly though steadily into Europe, where it remained until the collapse of the Roman Empire.

The emergence of Persia to its former position of dominance followed a short, erratic and highly confusing period of change but, following the defeat of the Lydian Kingdom (modern day western Turkey) Croesus, the conquering King, systematically conquered all territories up to the coast line of Asia Minor and beyond. A few years later the Persian power extended to Egypt, Thrace and Macedonia – thus it extended from the Caspian Sea to the Nile. By this time, the Persian Empire was larger than that of any in earlier times. It had major resources of manpower, had easy access to two major seas and was immensely wealthy. Although it had no indigenous precious metal resources, it had enmeshed the richest sources of Asia and offered a monopoly market for the gold derived from mines not in its own territory. Persian gold was derived from many sources, including some well established ones which it had previously acquired by conquest earlier in its formation, but it was the conquest of Egypt that provided the major continuity of supplies.

Croesus was himself a Hellenophile who, in order to conciliate Greek feelings regarding his important position in the Persian Kingdom, made princely offerings of gold and precious metals at the shrine of Apollo at Delphi. The riches of Lydia were based on the Greek River Pactolus, which transported huge quantities of gold down from Mount Tmolus in the Anatolian highlands. It was Lydian gold, associated with Croesus, which produced a great revolution along the Asia Minor shore line with the origin of the world's first true coinage (discussed further on p. 32 *et seq.*). Croesus introduced a pure gold coin and a pure silver coin, both marked with his royal device of the facing heads of a lion and a bull. Denominationally the gold was related to silver in the proportion of 1:10. Herodotus, the first 'Historian', wrote: 'The Lydians were the first men known to us who struck and used coins of gold and silver'. The first bi-metallic system of coinage had been devised.

The conquest of Lydia by Persia enabled the 'bi-metallic' system to be extended with only slight modifications when the gold '*daric*' and silver '*siglos*' (*shekels*) were introduced. This system was to last for hundreds of years and, such was the Persian wealth, the Kings had no qualms about the large personal fortunes built up by their subjects. As an indication of the immense Persian wealth, it is reported that when Alexander the Great seized the royal treasure at Susa he took over two million pounds of gold and silver in the form of ingots and a further 500,000 pounds of gold coin. Herodotus, who was familiar with the trappings of the Persian court, knew that the Persian Kings melted down precious metal received in tribute and kept the ingots in store. By the time of the collapse of the Roman Empire in the West, gold had been used for over 4,000 years for both trading and for storing wealth.

For mediaeval Christianity gold was the symbol of eternity and light. It had little material value, but it visually represented the religious value of the objects with which it was linked: gold crowns, altars, reliquaries and gospels. Byzantine theologians, Irish monks, German nuns – they all spent decades copying Bibles and painting illustrations and initials on golden backgrounds. The finished work was bound in gold and set with jewels. The evangelistaries were not just meant for kings and emperors. Every monastery and every cathedral needed a relic to document the presence and the mission of God. Likewise a costly gospel was visible proof of how greatly the word of the Lord was venerated.

Given man's craving for gold, it was inevitable that his inventive genius should be turned towards the search for producing gold by transmuting base minerals. Alchemists, as they were called, flourished in the East and in Arabia before the birth of Christ, and the varying mystery and magic which shrouded their researches continued to intrigue and excite man's imagination throughout the Middle Ages and beyond as knowledge of alchemical methods spread into Europe. The principal aim of the alchemist was to find the Philosopher's Stone, a mysterious agent which possessed the properties of producing gold and prolonging life. The possession of gold and eternal life have long dominated man's ambitions and the possession of the Stone, which would conveniently satisfy both these desires, became the over-riding obsession of the alchemist.

It is hardly surprising that the attention of charlatans and confidence tricksters should have been quickly turned towards alchemy and the spurious alchemist was often able to persuade his victims that the secret of the Philosopher's Stone was his, and could be theirs for a price. The easiest trick was to secrete a piece of real gold into a crucible, perhaps inside a cube of charcoal. The heat in the crucible would release the gold, astonishing and convincing the audience. In its heyday, alchemy was practised by Kings Heraclius I of Byzantine, James IV of Scotland and Charles II of England. While there is no evidence that alchemists succeeded in creating gold, there were permanent beneficial effects in that the research which took place did result in the emergence of first the art, and then the science, of chemistry.

By 1500 A.D., gold currencies were abundant in Europe. The reason for this was the opening up of the new world and the quest for El Dorado.

The Quest for El Dorado

Men's minds were turned to China, India and other mysterious parts of the Far East by the extraordinary journey undertaken by Marco Polo. His journal, published in 1295, gave an almost unbelievable account of the riches he had seen in China: 'in this province gold is found in the rivers and gold in bigger nuggets in the lakes and mountains'. The idea that strongly motivated Christopher Columbus was that these great sources of wealth could be reached by sailing westwards, across the Atlantic, a concept that he was eventually able to sell to King Ferdinand and Queen Isabella of Spain. His intention was to sail westward to Asia and establish a new trading route from Spain to the Orient. Although he did land in what became known as the New World, he appears to have assumed that the West Indies were the Orient, that the island of Hispaniola (currently Haiti and the Dominican Republic) was Japan, and that Cuba was the 'extreme edge' of the East'. These illusions may have been deliberately fostered in order to gain further support, for his royal employers were expecting results and Columbus was anxious to justify his theories.

Although not immediately successful, Columbus eventually did discover evidence of the existence of gold in the region, and this news quickly excited Spanish interest in new lands. Further expeditions set sail to colonise the territory already claimed for Spain and to uncover the wealth of gold rumoured to lie in the hinterland of central America. The instructions of King Ferdinand were plain: 'Get gold, humanely if you can, but at all hazards get gold'. Humanity was to play little part in the search.

Hernando Cortez first set sail for the Indies in 1504, where he established himself as an influential citizen of Cuba. In 1518 he sailed for the mainland, determined to find the lands of gold. He and his companions landed near Vera Cruz, and learned for the first time of the inland country of Mexico and of the Aztec civilisation there. The Aztec King Montezuma sent gifts and words of welcome to Cortez, and the Spaniards pressed on towards Mexico, anxious to lay their hands on the vast stores of gold which they now knew existed in the Aztec capital. Gold was of little value to the Aztecs except as decoration, and if the possession of it would placate the pale invaders, Montezuma was prepared to give it freely. But there was another reason why he allowed Cortez and his small force of men to advance unchecked on his capital: Aztec myth foretold the return to earth of the great god Quetzalcoatl. Montezuma was not prepared to take any chances. He was astute enough to realise that Cortez was merely the forerunner of a greater invasion and that the survival of his nation might well depend on his kind treatment of the new arrivals.

Montezuma's capital, Mexico-Tenochtitlan, astonished the Spanish when they eventually reached it. They were dazzled by the wealth of the island city and by the lavish gifts of gold and precious stones which Montezuma showered on them. Montezuma surrendered himself to Cortez, becoming a vassal of the king of Spain, and the Spanish triumph was complete. But when Montezuma's brother launched an attack on the Spaniards the King was killed in the fighting that followed and Cortez was forced to withdraw from the capital, taking his gold with him. But this was only a temporary setback to Spanish ambitions. By 1521 the Aztecs were finally conquered, and Mexico, with all its gold, became part of the Spanish empire.

Far to the south of Mexico, in the vertical landscape of Peru, lay the kingdom of the Incas. Rumours of the wealth of this country and of the immense stores of gold held by the king soon filtered northwards and reached the eager ears of the Spanish. An expedition led by Francisco Pizarro set out in 1526 to see whether the reports of this golden civilisation were true.

Pizarro's discoveries in Peru exceeded his wildest expectations. Gold abounded in the Inca cities, and their sun temples were covered in golden ornaments of the greatest beauty. The Incas worshipped the sun, and gold was inextricably linked with the god; gold was referred to as 'the tears wept by the sun'. Before long, the Inca King, Atahualpa, was captured by Pizarro at Caxamalca. In an attempt to secure his release, Atahualpa offered to give Pizarro the amount of gold objects, rather than gold bullion, which covered the floor of his apartment, an offer quickly accepted. The apartment was approximately seventeen feet broad by twenty-

two feet long, and it was agreed that it should be filled nine feet high.
Although the ransom was eventually paid, some six tonnes in weight, disputes between the Spaniards as to its division, combined with evidence of political uprising, led to the eventual garotting of Atahualpa. The gold of Atahualpa and the silver of Montezuma were delivered into the hands of Spain which, on the proceeds, rapidly evolved into the most powerful nation in Europe. This flow of gold, in unprecedented quantities, led to a prolonged inflation throughout the continent. It was a precise repetition of the price explosion which had shaken Alexander's empire eighteen hundred years earlier.

The Spanish were not content with the acquisition of the golden treasures of Mexico and Peru. Surely, among the dense jungles of Central and South America, other civilisations were waiting to be pillaged? Surely there were further golden cities to be sacked, new treasures to be found? Soon rumours spead of a fabulous golden city, standing near a lake, which offered wealth and riches far in excess of those previously found. This legendary city, known as El Dorado, had developed from stories which the Spaniards had heard about the existence of *el hombre dorado*, the gilded man – a king who, in some sort of ritual observance, covered himself in gold dust before bathing in a lake.

The legend consisted of three elements: a gilded man, a lake filled with gold as part of a religious ceremony (which historians have tried to identify with Lake Guatavita) and a city furnished with gold. Lake Guatavita, high in the mountains of Colombia, was one of the sacred lakes of the Muisca Indians, and it was here it was said that each new ruler went through a strange ceremony before taking office.

> 'They stripped the heir to his skin, and anointed him with a sticky earth on which they placed gold dust so that he was completely covered with this metal. They placed him on a raft . . . and at his feet they placed a great heap of gold and emeralds for him to offer to his god . . . The gilded Indian then made his offering, throwing out all the pile of gold in to the middle of the lake . . . From this ceremony came the celebrated name of El Dorado, the Gilded Man, which has cost so many lives.' (Juan Rodriguez Freyle, 1636)

Belief in this legend has kept searches carrying on ever since. Several kinds of expeditions have tried to conquer the lake with every kind of equipment from drills to mechanical drags and airlifts. Although the central zone remains untouched, many of these teams had partial success and picked up a few more gold items to add to the spoils until, in 1965, the Colombian Government brought Guatavita under legal protection as part of the nation's historical and cultural heritage.

The era of gold production that followed the discovery of the Americas was in all probability the greatest the world had witnessed to that time. The exploitation of mines by slave labour and the looting of palaces, temples, and graves in Central and South America resulted in an influx of gold that literally unbalanced the economic structure of Europe and disturbed its political makeup. From the discovery of America by Columbus in 1492 until 1600, more than 8,000,000 ounces (225,000 kilograms) of gold, or 35 per cent of world production, came from South America. South American mines, especially in Colombia, continued into the seventeenth and eighteenth centuries to account for 61 and 80 per cent, respectively, of world production; 48,000,000 ounces (1,350,000 kilograms) were mined in the eighteenth century. Russia became the leading producer in 1823, and for fourteen years contributed the bulk of the world supply.

The California Gold Rush and the Comstock Lode

During the second great era of expanding production (1850–75) more gold was produced in the world than in all the years since 1492, chiefly because of discoveries in California and Australia. The story of the California gold rush begins with the arrival in 1845 of James Marshall, a carpenter, in California and his subsequent employment with John A. Sutter. Marshall, a sawmill foreman, although difficult to get on with, was a good builder and was keen to complete his job of building a new sawmill. While his men dallied over breakfast he was already busy down at the river. Each day the men cut the mill race ditch deeper, and each night Marshall let the river rush through it to sweep away the debris. On the morning of January 24th, 1848 Marshall closed the sluice gate and waded down the drained muddy ditch, checking its depth. Suddenly something shiny caught his eye. What he found changed the course of American history. After putting two nuggets into his slouch hat he showed them to his workers who seemed unimpressed although one of them, Bigler, noted in his diary: 'This day some kind of mettle was found in the tail race that looks like gold, first discovered by James Martial, the Boss of the Mill.' Marshall decided to test these nuggets. He took a small rock and began pounding the yellow metal. It was soft and flattened out easily; this convinced him it was pure gold for if it was fools' gold it would have shattered,

On January 28, he rode wild-eyed and dripping wet into Sutter's fortress-like ranchero in the Central Valley and told his startled boss that he wanted to see him alone. 'My God!' yelled the almost paranoiac Marshall, when an unsuspecting clerk wandered in with some papers, 'didn't I tell you to lock the door!' He and Sutter bolted the door and shoved a wardrobe in front of it. But the word was already out. As Marshall's men worked the gold they spent their earnings in the store of Sam Brannan. Being a good businessman Brannan knew that any gold seekers would have to pass his store on their way to the gold fields. He stocked up with all the picks, shovels and other equipment the men might need. Then he went to San Francisco and walked the streets announcing the news and encouraging all to head for the hills and become rich overnight. Traces of gold had been found before in California, but had caused no stir. Now, in 1848, it was suddenly different. As one San Francisco newspaper lamented: 'The field is left half planted, the house half built, and everything neglected but the manufacture of shovels and pickaxes.' By fall, when the news reached eastern America, the whole country went wild. Gold-seeking 'Argonauts' from all over the world poured into California, multiplying its population sixteen-fold in four years, and doubling the world's gold output.

Although it took months for the news to spread, the subsequent gold rush dogged both Sutter and Marshall with misfortune. Sutter's workers abandoned their jobs in a thunderous quest for instant wealth in the hills, leaving thousands of dollars worth of wheat and hides to rot. His original dream of an agricultural empire ruined, Sutter tried to profit from the gold rush in a variety of ways, but never really succeeded. In the end he moved to Pennsylvania and died, heartbroken, in 1880.

Marshall became a folk hero to others but a failure to himself. Stubbornly claiming supernatural powers, he wandered through the hills looking for more gold, but with little success. Other prospectors nevertheless assumed he had the Midas touch, and infuriated Marshall by following him and digging wherever he dug. More and more embittered, he came to believe that all the gold in California was rightfully his. Rejected by most camps, Marshall moved to Kelsey, now an obscure ghost town just east of Coloma, where he lived on odd jobs and handouts until his death in 1885. Sam Brannan, however, made his fortune and ended up owning most of the valuable land in San Francisco.

Earlier discoveries of abundantly rich gold deposits had previously been monopolised by the State or at least heavily controlled by it. California was a new phenomenon: this latest El Dorado belonged to anyone who could get there and stake a claim, with its gold being passed into the direct possession of those who could find it. From 1851 to 1855 the yield of Californian gold was running at a rate of around 175,000 lbs. a year, with a peak of 200,000 lbs. in 1853. In these five years, California produced what it had taken ancient Rome half a century to win from North-west Spain. Set against current production in other countries at this period (for example, Russia) the gold of California poured out in a flood. In fact it amounted to about half of what was being produced all over the world at the time. In the nine years from 1848 to 1856 California produced 752,400 kilograms of gold. Figure 5.4 illustrates the trend of US gold output from 1820 onwards. The discovery of gold transformed California from an isolated pastoral community into a vigorous and prosperous territory with, in 1852, a population of over 250,000 people. In the next fifty years the pattern of immense mining migration, suddenly conceived urban settlement, and shifting capital was to be repeated on a major scale in several areas of the globe, each of them far removed from the traditional centres of mining enterprise.

By the late 1850s the Californian miners were restless, almost desperate. The easy pickings were gone. Suddenly in 1859 electrifying news came from across the Sierra: someone named Comstock had given his name to a fabulous silver strike. Ten years earlier, during the first hell-bent rush for Californian gold in 1849, thousands of eager gold hunters had tramped across the sagebrush desert that was then the western edge of Utah Territory. Some even stopped and picked up a gold nugget or two. But the land was parched and forbidding, provisions were short, and their minds were on California, where gold was said to be so plentiful it was bunching up in the streams. As the early prospector, Mark Twain, remarked about the area, 'No flowers grow here, and no green thing gladdens the eye. The birds that fly over the land carry their provisions with them.'

The real beginning of placer mining in what was to become Nevada was early in 1850 when a Mormon wagon train *en route* to California camped in Carson valley to rest and feed their animals. Several of the party, while prospecting along a nearby stream, found small specks of gold at a place named Devilsgate. Although the wagon train moved on word spread, thereby encouraging a small stream of prospectors to the area.

The prospectors, generally, weren't men of great intellectual ability. But two of them, Hosea and Ethan Allen Grosch, were well educated in mineralogy and came close to uncorking the secret that the mountain concealed – that it was a repository for the richest mass of silver the world has ever seen. Their letters have indicated that they had discovered a rich source of silver and gold. But their luck was bad. One day a tragic accident occurred; Hosea struck his foot with a pick. The wound became infected, and within two weeks he died. His brother, lonely and grief-stricken, set out across the Sierras for California. He started too late. He was caught in a blizzard. For four days and nights he struggled through blinding snow. But his hands and feet froze, and within two weeks he, too, was dead. With the Grosch brothers' death, the secret of their silver mine was lost.

Prospectors, however, still continued searching and by the Spring of 1858 ground in the diggings was at such a premium that two prospectors Peter O'Riley and Patrick McLaughlin decided to go to the head of a nearby canyon and began prospecting the slope of the mountain with a rocker (described in Chapter 1). Here, almost by chance, they stumbled on a small bonanza. Needing a better supply of water to work their rockers, they set to work opening the mouth of a tiny spring. On impulse they tossed a shovel load of dirt into their rocker. The results astounded them. The whole apron of their rocker was covered with a layer of bright and glittering gold. Every time they washed another bucketful through, the bottom was covered with gold. In no time at all, they were literally taking out gold by the pound. And there, in that little prospect hole, silver mining in the West was born. For, along with the gold that so elated the two Irishmen, there was a quantity of 'heavy black stuff' that puzzled them. They didn't know it, but it was silver. The Ophir mine alone, into which they had just accidentally dug, would yield $20 million before it was exhausted.

The stage was now set for the great events that were to occur which resulted in the local town, Virginia City, becoming the wealthiest town in the West, possibly in the entire world. Not realizing the importance of their find the two Irishmen gave up their claim to a one Henry T. Paige Comstock as he signed himself. Word of the silver strike spread like a contagious fever across the Mother Lode country in California. Over the Sierras they streamed, the knowledgeable men of the mines and the mills, along with every prospector and greenhorn afflicted with gold-lust.

As well as placer mining there was a considerable amount of mining of gold ore. Gold metallurgy really started to develop in the Black Hills of South Dakota and at the Homestake Mine around 1877. After crushing the ore was pulverised by stamps (a block that crushes ore in a stamp-mill) and the gold recovered by gravity and amalgamation. However, it was not long before the Black Hills became a centre for the next great leap in the technology of gold recovery. The genius of Charles W. Merrill and inventiveness of J. V. N. Dorr took the original cyanide process as a start and improved it by inventing and perfecting ancillary processes and equipment for separating sand from slimes, for dewatering, aerating, separating, cyanide leaching of sands, and pressure leaching of slimes, and for zinc dust precipitating of the gold from the cyanide solution. The simplicity of the design and operation of the new processes and new equipment led to great improvement in recovery and permitted substantial expansion in the size of equipment.

The impact of the gold and silver finds was felt far beyond 'Washoe', as the territory was called. Much of the wealth found its way to San Francisco, where an important financial and banking system grew up around the frenzied trading of Comstock shares; Virginia silver virtually underwrote the diversified growth of this new city-by-the-Golden Gate. The Comstock's factory-like atmosphere marked the first true industrial expansion in the West; as technological precedent and prototype, its influence was pervasive. And Washoe's sudden growth caused ripples in Washington, where Lincoln's government, drained by the Civil War and anxious for support in passing the Emancipation Proclamation, hurried the territory into statehood in 1864, renaming it 'Nevada'. However by 1880 the Comstock Lode had been substantially worked out.

Although being only some 4 miles long the Comstock Lode was incredibly rich. Following its discovery in 1859 it was America's primary source of gold for twenty years. Apart from the discovery, no new developments occurred in the USA until the 1890s. In the State of Colorado, gold was discovered in 1890 in Cripple Creek: the yield was up to 19 ounces a ton and this earned the discoverer, William Stratton, $125 millions in 10 years. In Alaska, an almost unpopulated dependency of the USA, the famous gold rush was sparked off by a few Scandinavians, but the gold was alluvial and quickly exhausted. Alaskan gold discovered in 1898 only acquired importance between 1900 and 1906, then fell off and finally dropped drastically.

Gold Discoveries in Australia, Canada and South Africa

In Australia it was an immigrant, E. H. Hargreaves, after having already worked in California, who prospected areas which geologists had pointed to, in 1847, as likely to contain gold. Work first began in 1851 in the Northern Bathurst region. Gold was found in the Blue Mountains at Sommer Hill and then in the MacQuairie River. After August 1851, strikes were made in the south, and in 1852 near Adelaide. The low level of population, the scattered nature of the mines and the recent development of Australia's sheep farming made the impact of the discoveries less dramatic than in California. The discoveries were, nonetheless, undeniably important in the six years between 1851 and 1856 500,000 kilograms of gold were produced. Gold production rose from 650,000 ounces a year between 1831 and 1840 to 6,300,000 ounces between 1851 and 1860 and then fell slowly back to 5,200,000 ounces between 1881 and 1890.

In the Klondike area of Canada along the River Yukon, which flowed down to the Pacific through Alaska, there had been talk of gold since 1886, but it was only in 1896 that gold bearing sands were discovered on the scale of the Californian discoveries of 1848–1850. In 1895 there had been no human habitation at the confluence of the Yukon and Klondike rivers. By 1897, Dawson City had sprung up to provide a supply centre for the prospectors. The prospector's problems here were not the same as those in earlier gold discoveries. Although gold could be washed from rivers and streams, its greatest abundance lay in a bed of gravel up to about 15 feet thick that was covered by a great deposit (up to 30 feet thick) of black and boggy soil that was, for most of the year, deep frozen. Once this problem was overcome, however, the yield was extremely high. An output of 80,000 ounces in 1899 leapt to 720,000 ounces in 1900.

The first great riches found in South Africa were in fact not gold but diamonds. The diamond fields at Kimberley on the banks of the Vaal River were discovered in 1867. They attracted, from all over the world, men who had previously been lured by gold. Now they built up diamond fortunes that would later enable them to participate in the next scramble for gold. They included figures such as Cecil Rhodes and his partner Charles Rudd, J. B. Robinson, Barney Barnato and George Albur, who were to establish the mining finance houses to nurture the South African gold mines.

There were a number of claimants to the discovery of gold on the Witwatersrand in South Africa, but it is now generally accepted that an itinerant handyman and prospector named George Harrison finally stumbled on the exposed, rich Main Reef Leader in 1886. George Harrison made the momentous discovery on the widow Oosthuizen's farm 'Langlaagte', a few miles west of where the city of Johannesburg stands today, on the northern rim of a geological basin that was once an inland sea.

A farmer wrote a letter on 9th June, 1886, announcing the discovery to the Transvaal Republic's President, Stephanus Johannes Paul Kruger. Harrison was then asked by Republican officials to make an affidavit as to what precisely he had found on the Witwatersrand. His affidavit stated simply, '. . . I think I have found a payable gold-field'. He had not exaggerated. Soon afterwards, the president's commissioners, C. Johannes Jouvert and Johann Rissik, were sent to control a mining camp on the site and to proclaim Johannesburg. Harrison, who sold his 'discoverer's claim' for £10 sterling, wandered off the historical stage almost as soon as he had come upon it.

Penniless prospectors and adventurers from all parts of the world, and skilled financiers with experiences of mining in the flourishing diamond fields of Kimberley, flocked to the area. In 1890 only 440,000 ounces were produced, the real stimulus coming later with production rising to 3,638,000 ounces in 1899. The mining camp of Johannesburg became a city. At the turn of the century a young war correspondent, Winston Spencer Churchill, came upon the scene and wrote:

> 'The whole crest of the Rand ridge was fringed with factory chimneys. We had marched nearly 800 kilometres through a country which, though full of promise, seemed to European eyes desolate and cold, and now we turned a corner suddenly and there before us sprang the evidence of wealth, manufacture, and bustling civilization.'

Gold had brought quickly into being an important mining and business centre – one of the few major industrial areas in the world at that time which had no access to a waterway. For a long time gold mining was confined to the 121 km belt known simply as 'The Reef'. This consisted, in fact, of three goldfields: the Central Rand, the West Rand and the East Rand, on which towns such as Johannesburg, Germiston, Roodepoort, Krugersdorp, Randfontein, Boksburg, Brakpan, Benomi, Springs and Nigel were established. Mining houses, using the talents of financiers, administrators, engineers and scientists, had taken over from carefree prospectors in the search for gold, but the hunt remained as keen as ever. Figure 5.3 (p. 137) illustrates the geographical location of South Africa's goldfields.

The Far West Rand gold-field was discovered at the turn of the century, but shaft sinking efforts proved abortive because of the presence of large quantities of underground water. In 1931, a German geophysicist,

Dr. Rudolf Krahmann, backed by Gold Fields of South Africa, set out to trace the 'lost' Reef to the west of Witwatersrand by using a magnetometer to locate the magnetic shale-beds associated with the gold-bearing ore. Geologists, engineers and company officials studied the magnetometer readings and set up their drilling rigs in search of the elusive Main Reef. Below thick layers of ancient lava they struck the reef now known as the Ventersdorp Contact Reef. Later the Elsburg Reef and the Carbon Leader were discovered.

As a result of the continuous worldwide exploration, the world's stock of gold rose dramatically. The yield in the sixteenth century may have accounted for about 1.5 per cent of all gold newly obtained from that time onwards. The seventeenth century may have produced fractionally more. The eighteenth century could probably claim nearly 4 per cent, and the first half of the nineteenth century about 2 per cent. But the estimate of gold produced from 1850 to 1900 – about 23,000,000 lb – is over 20 per cent. These fifty years yielded over twice as much gold as the two and a half centuries after Columbus' discovery of America. The figure might even have been higher if research into the amalgamation and refinement of gold from crushed ore had been undertaken a little earlier. For until 1891 something like a third of the gold from crushed ore like that of the Witwatersrand mines was being wasted by imperfect chemical methods of separation, and was literally flowing away in sludge and slime. The McArthur-Forrest cyanide process (*see Chapter 1*) then reduced the wastage to a mere 5 per cent or so.

This new process was, indeed, the answer to a whole new set of problems. Until 1850 the great proportion of gold obtained from the earth was washed gold. Even in the Californian, Australian and Klondike gold-rushes, huge quantities of gold were accounted for by the simple and age-old method of washing. But in California itself the disintegration and washing of massive gravel terraces had quickly pointed the way to the need for the heavy hydraulic power that Roman miners had done their best to devise in Spain. From hydraulic power for washing out the terraces to other forms of power, applied in particular to the crushing of ore, was a short step in a period when ample capital could be attracted by the expectation of such rich dividends. The general change in the methods of gold-mining from 1850 onwards, caused jointly by the lapsing of royal monopolies, by the influx of private capital, and by the development of machinery, is reflected in the statistics. From about 1850 to 1875 some 90 per cent of newly produced gold was alluvial, the remaining 10 per cent being mined from quartz veins. By 1890 the proportion of alluvial gold had been cut to 45 per cent and, significantly, the gold from deep-lying conglomerate beds or 'blankets' (so called from their resemblance, weathered brown at the surface, to Dutch burnt-sugar and almond-studded sweets of that name) had begun to make its mark with 8 per cent. By 1904 the figures were, respectively, 18 per cent, 60 per cent and 22 per cent, and by 1929 they were to be 8 per cent, 39 per cent and 53 per cent.

The Development of Gold Coinage

The invention of coins was not the inspiration of a moment, but the result of a long development. Thousands of years earlier there had already been metal money, predominantly in the form of axes and rings of copper, bronze, silver and gold. Above all, this money was a magical symbol of power and worth. The axe was an attribute of Kings and Gods, the ring warded off demons, inspired superhuman strength, and guaranteed eternal faith. The beauty of the precious metals used for making axes and rings (especially gold) strengthened their divine energy. The more treasures a man owned, the greater his magical might. The symbolic metallic money was, hence, not to be spent, but to be gathered and hoarded. The credit for the invention of coins must be given to the Chinese, who, about 1000 B.C., manufactured gold coins in the shape of knives and spades, in imitation, one may presume, of the forms of money which preceded them. The first European rulers recorded as having struck coins were the Kings of Lydia. Herodotus, the Greek historian, writing about 430 B.C., acknowledges the Lydians as 'the first people we know of to strike coins of gold and silver'. Historians are generally agreed that these first primitive coins were made about 650 B.C. during the reign of King Ardys (652–615 B.C.). They were made from a natural alloy of gold and silver found in Asia Minor, known as electrum. Gold and silver had been used in trade for centuries before this but each nugget or ingot had to be tested and weighed every time it changed hands. A new type of Lydian coin appeared about 600 B.C. under King Alyattes (610–561 B.C.). The heaviest coins, which weighed 168 grains, are called staters, meaning 'weigher' or norm of value. These coins, again made of electrum, had the very great weakness of varying considerably in the amalgam of the two metals.

The first large scale issue of gold coins occurred under King Croesus of Lydia (561–546 B.C.). The wealth of Lydia was so great that one recorded gift of bullion and precious metal ornaments which Croesus made to the shrine of Apollo at Delphi (home of the Delphic oracle) has been calculated to have contained nearly four tons of gold. Croesus' coins, minted at Sardis, were staters containing about 98 per cent gold, as close to the pure metal as possible with the refining methods then known, thereby resolving the problems associated with the use of electrum. The Lydian Empire was toppled by Cyrus of Persia in 546 B.C. and the Lydian issues came to an end. Although the Persians took over the mint at Sardis, Cyrus did not issue any coins during his reign. His successor, Darius the Great (521–486 B.C.) however, struck enormous quantities of coins in both gold and silver. The standard gold piece of the Persians was the *daric*. This name and the name of the ruler himself are derived from the Persian word dara, meaning king. Like Croesus, Darius used nearly pure gold for his coinage. Experience had shown, however, that the addition of a small amount of alloy resulted in harder, longer lasting coins, and about 3 per cent of base metal was deliberately added.

It was not long before all the Greek states followed Lydia in manufacturing coins, many of which still exist and are master-pieces of beauty. Philip II of Macedonia, who had gold mines in his kingdom, produced great quantities of gold staters – or 'philips' – which became acceptable throughout the Greek world. His son, Alexander the Great, who seized the golden hoard of Darius III, augmented the existing supplies so that coins struck by Greeks reached Rome, and became the chief gold currency of the Republic. According to Pliny the Elder, the first coins produced by Rome were issued by Servius Tullius, one of the early kings. Augustus was the first Roman to strike a golden currency, known as the *aureus*. The Roman Republic, however, soon acquired the habit of debasing the currency. Under the Empire, the coinage was invariably debased by bad Emperors, and restored by good ones. The emperors following Augustus continued to strike the *aureus*, but under Nero (A.D. 54–68) it was permanently reduced in weight, though the purity of the gold was maintained. Nero decreed that 45 *aurei* were to be struck from a pound of metal, rather than the previous 42 pieces. The Roman mint was situated in the Temple of June Moneta. This has caused some people to believe that our word 'money' was derived from the name of this temple. Others believe that this suffix was given to this particular temple of Juno because of the 'moneta' which was manufactured in it.

An increasing number of barbarian attacks along the frontiers plus other difficulties forced Emperor Constantine the Great (307–37 A.D.) to transfer his court from Rome to Byzantium, which he renamed Constantinople (now Istanbul). This had the effect of dividing the Roman world into two parts: the East administered from Constantinople, the West by an assistant emperor in Italy. Constantine the Great reformed the coinage, replacing the gold *aureus* which had gradually declined in weight with a new denomination, the gold *solidus*, actually the *solidus nummus*, or 'thick coin'. Lighter than the *aureus*, the *solidus* was produced at the rate of 72 from a pound of pure metal, rather than the 60 pieces then in effect for the *aureus.*

In Western Europe, the Roman frontiers collapsed in the fifth Century and tribes of barbarians spread over the former Roman provinces. The Western Roman Empire ended in 476 A.D., when Romulus Augustulus (475–76 A.D.) was defeated by the barbarian Odoacer. Having no coinage of their own, the barbarians took over the Roman currency and mints, striking coins whose designs copied the appearance of the issues of the remaining Eastern Imperial mints.

In the Eastern Empire, the gold *solidus* introduced by Constantine the Great was the standard gold coin of the Byzantines. A smaller one-third *solidus* piece called the *tremissis*, or *triens*, was, however, added and first struck at Constantinople in 383 A.D. Although the Byzantines drastically reduced their bronze coins in size over the years, the gold pieces were kept absolutely stable for centuries and, in the absence of Western gold coinage, they circulated throughout Europe, where they were known as *bezants*. The weight of the solidus, maintained at the same level for seven centuries, was equal to 24 carats, an ancient unit of weight measurement. The carat also came to be used as a standard of fineness. The word 'carat' comes from a Greek word meaning a carob bean; these little seeds were originally used as weights. Anything other than pure gold that went into the *solidus* became in effect a measure of its debasement. Thus gold that is said to be 23 carats fine consists of 23 parts of pure gold to 1 part of alloy. Pure, unalloyed gold is 24 carats.

During the Middle Ages, a multiplicity of mints sprang up throughout Europe. Coins were struck by kings, barons, chartered towns, and by the Church. Many of these mints struck coins in imitation of those which had existed in Rome before the fall of the Empire. Thus Roman coins, such as the famous *denarius*, survived, at least in name, for centuries. The standard gold coin of Florence, issued in the thirteenth century, derived its name from its design, a lily, which was the emblem of Florence. The Italian term *fiorino d'oro* means 'little flower of gold'. The anglicised form of the name is florin, and this name and that of the ducat have been applied to similar gold coins of many other nations.

French regal gold coinage began in 1266 during the reign of Louis IX (1226-70). This first gold issue was called an *ecu d'or*, literally a 'shield of gold'. This coin carried the *fleur-de-lis* coat-of-arms. The *fleur de lis* (the three flowers of the white lily joined together, one erect, the other two curving outwards) was first used as an emblem by King Louis VII in 1147.

Edward III (1327-77) was the first English monarch to institute a regular gold coinage. Gold coins had been struck intermittently before this – the Anglo Saxons issued gold *tremisses* early in the seventh century, and Henry III struck a gold penny in the middle of the thirteenth century – but Edward made gold an integral part of the nation's monetary system. England's growing commerce, especially in the wool trade with Flanders, created a need for higher denomination coins and gold was the logical answer. The first issue, a florin, its half and quarter, came in 1344, but the actual weights were below their official value in relation to the silver pieces then in circulation. Not readily accepted, the first issue was soon demonetized and replaced by the heavier gold noble. Henry VII (1484-1509), taking a personal interest in the coinage, created a special commission to oversee the striking of a new gold piece originally called a double royal, but later changed to the sovereign.

Machine made coinage was first tried in England during the reign of Elizabeth I (1558-1603). The coinage in circulation at Elizabeth's accession was badly debased and often clipped as well. The Queen remedied the situation by devaluing the existing underweight coins and introducing the first machine-produced milled coinage in 1561. This was however, later abandoned and it was left to Charles II (1660–1685) to produce the first quality controlled gold coins. The first of the new coins was a twenty shilling piece dated 1663. Much of the gold for the new coinage was brought from Guinea in Africa (now Ghana, 'land of gold'). The specimens made of Guinea gold carried a small elephant device on their obverse and, from the source of the metal, received the name 'guinea'. The gold was imported by the 'Company of Royal Adventurers of England trading in Africa'. Its charter granted the company the privilege of having coins bearing its badge struck from the precious metal it brought back to England. Coins with values of five, two, and one-half guineas were also struck. The five guinea piece, containing nearly an ounce and a half (avoirdupois) of 22 carat gold, is the heaviest English gold piece ever circulated.

The discovery of the New World by Columbus in 1492 and the subsequent exploitation of its wealth of gold gave a new impetus to gold coinage in the early modern period. Much of the gold of the New World was turned by Spain into coin by a chain of mints stretching from Mexico down to Lima and Potosi in Peru. The gold coin, the *escudo*, was produced as a single piece, and very frequently a double *escudo* piece. The double *escudo* piece passed into our language as a doubloon. The common designs of these and even larger pieces of four and eight *escudos* were the Pillars of Hercules and the Spanish arms of Leon and Castille.

During the fourteenth, fifteenth and sixteenth centuries, Europe experienced a re-awakening in outlook that was soon mirrored in literature and in the arts. Painters and sculptors turned from the harsh, stereo-typed Gothic forms to natural, realistic representations. Skilled artists developed the technique of making beautiful, high-relief portrait medallions, then turned their talents to coins.

In the kingdom of France itself, the first monarch to place his portrait on a gold coin was Henry II (1547–59). His head appeared in 1549 on the standard *ecu d'or* as well as on a new heavier coin, the *Henri d'or* introduced the next year and named after him. The new gold piece was also minted in half and double units. King Louis XIII (1601–43) completely overhauled the French coinage. Most of the coins previously in circulation were of less than full value from having had bits of metal 'clipped' off. Hammered coins with their irregular edges invited clipping and the only remedy lay in striking perfectly round coins, something possible to do only with machinery. By 1640, the Paris mint was won over to the production of machine-struck coins and Louis introduced a whole new series of coins in all metals. The smallest of Louis' new gold pieces was the same weight as the old *ecu d'or*, but only 22 carats instead of 23. Louis also struck coins of double this weight which, like the small piece, carried his portrait. The larger coin was called the *Louis d'or*, after its design; the smaller the *demi-Louis*. The *Louis d'or* was the same weight and fineness as the Spanish gold *pistole*, a coin then known all over Europe.

In Europe, during the 1700s, the general trend was for the larger, more powerful countries, to expand by absorbing the smaller independent states. The *ducat*, along with its multiple and fractional pieces, was the most important gold coin of the century. It was the primary gold piece in the Austro-Hungarian lands of the Holy Roman Emperors, in Germany, Scandinavia, the Netherlands and northern Italy (where it was known as the *zecchino*).

Under Louis XV (1715-74), abuse of royal power created mounting unrest. Louis XVI (1774-92) tried to meet the demands of his people, but revolution broke out and, in 1791, the King was forced to accept a constitution which reduced him to a figurehead. During his reign, Louis XVI had continued the denominations of the preceding reigns. The 1790 *Louis d'or* is one of the last issues of the French monarchy. Following the creation of the Directory, Napoleon Bonaparte, in 1802, became the First Consul for Life. Napoleon's first gold pieces were 20 and 40 franc values designed by Pierre Joseph Tiolier.

In terms of size, among the smallest gold coins ever struck are the gold *fanams* of Southern India which were minted in the early 1800s. They measured less than one-eighth of an inch (or 3 mm.) across. The absence of coinage material other than gold and the need for small denominations made the tiny size mandatory. On the other end of the scale are the 100 *zecchinis*, the largest gold coins ever produced in Europe. Issued by the Republic of Venice in the 18th century, the *zecchinis* measured nearly four inches (about 10 cm.) across. The coin has the figure of a standing Christ holding an orb which represents the world.

Although it cannot claim the distinction of being the largest, the 200 gold *Mohur* (weighing some 68.73 ounces), also from India, can boast of being the highest denomination coin that has ever been struck. Minted in the seventeenth century during the reign of Shah Jehan, who was also famed for having built the Taj Mahal, the coin had a nominal value of $1,500 in the 1960s.

During the nineteenth century, most of the commercial nations of the world, including America and India, adopted the gold standard as the basis of their monetary system (*see Chapter 3*) so that the expanding world economy of the period was served by what, in practice, amounted to an international monetary system. It did not matter that the chief coins of the different nations were of different weights so long as they were gold, and could be exchanged freely. This fact is well illustrated by a gold coin struck in 1874 by the US Mint. It is inscribed with the following six values: $10, £2 1s. 1d., Marken 41.90, Kronen 37.33, Gulden 20.73, Francs 51.81.

One of the best known and dominant gold coins is the British sovereign, first introduced after the Napoleonic Wars in 1816 with its designs by Pistrucci of the royal portrait and St. George and the Dragon. Much of the coinage was provided by the Royal Mint in London, but later in the century, following the great gold discoveries in Australia and South

Africa, the sovereign was struck at branch mints set up in these countires, distinguished only by a small initial letter on the reverse, for example 'M' for Melbourne, and so on.

The greatest age of gold coinage was the hundred-year period between the downfall of Napoleon and the outbreak of World War I. Rich gold strikes were made during those years, the world was comparatively peaceful and commerce flourished, and gold coins were in circulation nearly everywhere. The 1970s have resulted in very significant interest in the Krugerrand and this is discussed further in Chapter 6. The Appendix on p. 239 provides a detailed history of the world's gold coins, where they came from and the years they were produced.

The Development of US Gold Coinage

The earliest settlers of British North America had little need for coinage, making do instead by bartering skins, tobacco, beads, etc. But as the population grew and trade opened up, so did the need for coined money. Traders arriving from Europe demanded coined money for the manufactured goods they brought in. The colonists accepted and used nearly all foreign coins that they were offered. Consequently they accepted and made payment in the French Louis, English guineas, German thalers, Dutch ducats, Portuguese moidores and, most predominantly of all, *peso a ocho reales* or pieces of eight, which were Spanish milled dollars.

When the government of the newly constituted USA took up the question of coinage, it settled upon the Spanish dollar for its monetary unit. Thomas Jefferson noted that 'the Spanish dollar is a known coin and the most familiar of all to the mind of the people'. The name dollar originated in Bohemia where a rich, new silver mine was opened in 1516 at Joachimsthal (Joachim's Valley). The large silver coins soon being struck there were called *thalers* from the source of the metal. The name was taken into other languages where it became *daler* in Dutch, dollar in English. The English name, the dollar, was applied to the Spanish *peso* or piece of eight (*reales*) which were the silver coins most widely used in the British North American colonies.

US national coinage was established on the decimal system (progressions of ten), however, rather than the Spanish system based on eight. The first coinage act of April 2, 1792, said in part: 'The money of account of the United States should be expressed in dollars or units, *dismes* [original spelling of dime] or tenths, cents or hundredths...'. This same act provided for coins of ten denominations ranging from a copper half cent to a ten dollar gold piece called an eagle.

The same 1792 Act of Congress provided for the construction and operation of a Mint at Philadelphia, then the seat of government. The Mint became the first building erected by the US Government, and the first official coins were minted in 1794 with gold eagles and half eagles ($5) being struck in 1795. The other gold denomination authorised, a quarter eagle ($2.50) was first produced in the following year.

No mark of value was used on any of these early US gold pieces. At that time, weight and fineness counted for more than officially stated denominations. The actual value of gold and silver coins fluctuated in proportion to the market value of the metals themselves. The US gold pieces were valued at 15 to 1 in relation to silver but, in Europe, the ratio had risen to 15.75 to 1. This made it profitable to exchange silver coins for the newly minted US gold pieces, which were then shipped to Europe, melted down and sold for a 5 per cent profit. Consequently, US gold coins were rarely seen in general circulation after 1800 and minting of the eagle was discontinued for thirty years after 1804.

A new coinage Law, passed on June 28, 1834, revalued gold at 16 to 1 in relation to silver. This was accomplished by reducing the purity of gold from 22 carats (0.9166) to 0.900 fine. With the new ratio, it was no longer profitable to melt coins. The quarter eagle and half eagle were redesigned, and large quantities were struck. Following the Californian gold rush of 1849 the flood of gold onto the world's metal markets completely changed the relative prices of gold and silver. The price of silver in relation to gold increased to the point where it became worthwhile to melt silver coins for their bullion value. To take the place of the rapidly disappearing silver pieces, gold dollars were authorised and struck in 1849. A larger denomination, the $20 gold piece, the double eagle, was also authorised, but mintage did not begin until 1850. The great amount of gold arriving at the mint had to be converted into coins, and it was cheaper and easier to produce one large denomination coin equal in value to several smaller coins. In 84 years of production, nearly 175 million gold $20 pieces were minted. The gold dollar was minted until 1889 and the double eagle until 1933.

At the outbreak of the Civil War, coins became scarce. During this emergency the USA in effect went off the gold standard when Congress authorised the issue of $450 million of US notes – the so-called 'greenbacks'. These notes were issued with the understanding that they would be redeemed and returned as soon as the emergency ended. But the greenbacks never were actually retired; under provisions included in the Redemption Law of 1879 they were re-issued as fast as they were redeemed and therefore allowed to circulate indefinitely. The South was on a pure fiat money basis from the beginning of the Civil War and the Confederate paper currency rapidly declined in value eventually becoming worthless. During this time period all coins disappeared, a classical example of Gresham's law, whereby the issue of bad money drives out good money.

During the remainder of the nineteenth century, the US Treasury experienced recurring difficulties trying to maintain full convertibility for the various issues of government paper that had come into existence. In addition to the greenbacks, there were then silver certificates and Treasury notes of 1890. The Gold Standard act of 1900 finally made it official that all forms of US currency were to be maintained at full parity with gold and to be fully convertible into gold coin or bullion. Table 2.1 details the major gold coins issued between 1838 and 1933. Table 2.2 gives the gold contents of these coins.

Table 2.1. Regular US Gold Coinage, 1838–1933.

$20	Double Eagle	Liberty Head	1849–1907
$20	Double Eagle	St. Gaudens Type	1907–1932
$10	Eagle	Liberty Head	1838–1907
$10	Eagle	Indian Head	1907–1933
$5	Half Eagle	Liberty Head	1839–1908
$5	Half Eagle	Indian Head	1908–1929
$2½	Quarter Eagle	Liberty Head	1840–1907
$2½	Quarter Eagle	Indian Head	1908–1929
$1	Gold Dollar	Liberty Head	1849–1854
$1	Gold Dollar	Indian Head	1854–1889

Table 2.2. Weights of US Gold Coins, 1838–1933

Coin	Grains Standard	Grams Standard	Grains Pure	Troy oz. Pure	Troy oz. (Nominal)
$20	516	33.4370	464.4	.9675	1.00
$10	258	16.7185	232.2	.48375	.50
$5	129	8.3592	116.1	.241874	.25
$2½	64.5	4.1796	58.05	.1209375	.125
$1	25.8	1.6718	23.22	.04837	.05

Notes:
1. All coins .900 fine.
2. Pure means unalloyed gold.
3. Standard is total weight of coin.
4. Difference between nominal value of gold dollar and actual gold content by weight is due to the original seigniorage granted the Treasury by the Law of 1837, under which the statutory troy ounce worth $20.67 in bullion was exchanged for $20 in coin. This small "profit" was allowed to offset minting expenses.

The period 1900 until 1932 is sometimes referred to as the US golden age. This golden age, which lasted some 32 years, ended when, in the midst of a national crisis, the worst since 1860, President Franklin Roosevelt issued, on March 5, 1933, a 'Proclamation of National Emergency'. This temporarily suspended all banking operations, ended the minting of gold coins, prohibited all further transactions in gold coin or gold certificates, and forced the surrender of most privately held gold coin and bullion to the Federal Reserve System by threatening imprisonment for the holders on the grounds of 'hoarding'.

An Emergency Banking Act, passed on March 9, 1933, gave Congressional sanction to the measures taken by the President, and under this Act, on April 5, 1933, regulations were issued permitting the limited use of refined gold for certain industrial and technical purposes and allowing the private possession of 'rare and unusual gold coins of special value to collectors'. There were further technical changes and orders regarding gold issued during the remainder of 1933, and in January 1934 the passage of the Gold Reserve Act formally ended the era of gold coinage in the United States, an era that had lasted (with occasional interruptions) for 96 years. Gold coins were demonetised, gold was nationalised and made a government monopoly, and the 'defined' dollar was devalued from $20.67 to $35.00 per ounce troy of gold.

In early 1961, prohibitions on gold ownership were extended further. As discussed in Chapter 3 gold had been selling in London at a price exceeding the US official price of $35 per ounce. The US Treasury had been selling gold to depress its price through a complex arrangement with the Bank of England. US gold stocks were at a 21-year low and Americans were reported to be purchasing large quantities of gold overseas. In an effort to slow the official gold drain, President Eisenhower, just before leaving office, redefined the scope of gold prohibition to persons 'subject to the jurisdiction of the United States'. Effective on June 1, 1961, the prohibition of gold ownership pertained to persons and business organisations owned or controlled by persons who were citizens of, residents of, or domiciled in the USA, wherever located, and those conducting business under the laws of the USA.

During the entire 1934 to 1974 period, legitimate and customary uses of gold in an industry, profession, or art had to be licensed in accordance with the provisions of the Gold Regulations published by the Treasury. The ownership of shares of stock in gold mining companies was not prohibited. Americans were permitted to own any amount of gold jewellery or fabricated gold. And the acquisition, holding and exchange of gold coins of recognised value to collectors was permitted without a licence.

The President was given the authority to eliminate gold prohibition on September 21, 1973 with an amendment to the Par Value Modification Act, but this authority was never exercised. Freedom to own gold was restored on December 31, 1974 under an amendment to the International Development Association Bill, approved by Congress on August 14, 1974. Despite the objections of Treasury Secretary William Simon and Federal Reserve Board Chairman Arthur Burns to setting a date certain for the repeal of the gold restrictions, the bill provided for automatic removal of restrictions on private gold ownership.

From January 1974 it had again become legal for US residents to import gold coins bearing a pre-1960 mint date. This opened the way for imports of restrikes from Mexico and Austria but the law still precluded the importation of Krugerrands. That restriction, as mentioned above, was removed upon full legalisation of gold ownership in the USA on December 31, 1974. With no new issues of US gold coins, the Krugerrand quickly emerged thereafter as a popular medium of investment for the US public.

Chapter Three

THE GOLD STANDARD:
ITS HISTORICAL ROLE AND
RELEVANCE FOR THE FUTURE

Introduction

What is a Gold Standard?

The international monetary system which was generally adopted among major industrialised countries before the First World War is referred to as 'the Gold Standard'. On a full gold standard the money of the country concerned, be it dollars, pounds, francs or whatever, consists of a fixed weight of gold of a definite fineness; the price of gold is fixed by law; and there is complete freedom to buy or to sell gold, to import it or to export it. When Great Britain was on the gold standard before the First World War the Bank of England was legally obliged to buy gold at £3 17s. 9d. and to sell gold at £3 17s. 10½d. per standard ounce. This gold was 11/12 fine.

A gold standard really performs two functions, to a certain extent separate and distinct. These functions are both internal and external to the economy. On the full gold standard, gold coins are in circulation and on demand bank notes can be exchanged for gold. This means that an adequate reserve of gold must be kept by the Central Bank to meet the demand. In Great Britain the Bank Charter Act of 1844 permitted a small fiduciary note issue, but the rest of the note issue required a full gold backing. The amount of the note issue in the country then depended on its stock of gold. If the stock increased the quantity of currency could be expanded; if it declined the amount of currency had to be reduced. On the gold standard the volume of currency is then directly and rigidly controlled; the volume of bank deposits is dependent on the amount of cash held by the commercial banks, and so this kind of money also is indirectly controlled. The gold standard's internal function is to control growth of the domestic money supply. Thus if the fiduciary issue (if there is any) is fixed, the volume of a country's currency depends on its stock of gold and so it is impossible for an over issue of bank notes to occur thereby ruling out the possibility of inflation or hyperinflation.

In the international field the gold standard provides stability of exchange rates between different currencies, for the exchange rate cannot fluctuate beyond the limits of the gold points (discussed below). In addition in international dealings the gold standard provides, assuming a country keeps to the rules, an automatic solution to its balance of payments adjustment problem. Certain preconditions are necessary for the successful operation of a gold standard. First, since in theory gold would be the only form of internationally acceptable money, domestic monies must be freely convertible into gold in order to permit international trade. Second, individuals must have complete freedom to import and export gold. Third, Central Banks must stand ready to buy and sell unlimited amounts of their own currency at a fixed gold price. Finally, each country's gold stock has to form the reserve base of its domestic money stock, thereby ensuring that domestic financial policies are dependent on any gold inflows and outflows.

Each country in the system would have a fixed gold price for its currency. Exchange rates between different currencies would therefore be automatically determined if for example, the Bank of England fixed the price of £1 at 112.982 grains of pure gold, whilst the US authorities set $1 = 23.2 grains, then the sterling-dollar exchange rate would be £1 = $4.87 (i.e. 112.982/23.2). This would be known as the mint par rate. On either side of this value would be a gold import point and a gold export point. The gold import point is the exchange rate above which gold flows into a country and the gold export point is the exchange rate below which gold flows out of a country. The precise locations of these points would be determined by the cost of shipping and insuring gold bullion. Suppose the costs of shipping and insuring gold were 1 per cent, i.e. 1 cent per dollar, then the gold import point would be $4.92, the gold export point $4.82. The sterling-dollar exchange rate would be kept strictly within these limits by the profit seeking activities of arbitrageurs. At any rate below $4.82 gold flows out and at any rate above $4.92 gold flows in. An example will help to illustrate how this would be accomplished.

Assume for instance that an excess supply of sterling temporarily pushes the exchange rate to $4.70 in London. Under such circumstances, it would profit an American arbitrageur to convert $4.70 into £1, use the £1 to purchase 112.892 grains of gold from the Bank of England, then resell the gold to the US treasury at the fixed price $1 = 23.2 grains. For his 112.982 grains, the arbitrageur would therefore receive $4.87, which yields him a net profit of 12 cents on the transaction (17 cents less 5 cents costs incurred in transporting the gold from London to New York). The incentive for arbitrageurs to shift gold from London to New York remains as long as the transport costs involved are less than the price differential between the two markets. The increased demand for sterling forces up the sterling/dollar exchange rate until sterling reaches $4.82. At this stage the incentive to shift funds from the United States to England is removed. ($4.87 minus 5 cents gives $4.82.) Similarly $4.87 plus 5 cents gives a rate of $4.92 at which gold imports would be unprofitable.

Thus the authorities do not intervene in the foreign-exchange market, they merely stand ready to buy and sell domestic money at a fixed gold price; it is the profit-seeking activities of gold arbitrageurs which maintains the exchange rate between the gold import and export points. Since a single gold reserve had to serve both internal and external needs, the export or import of gold to meet the needs of the balance of payments will affect the internal situation. An export of gold reduces the gold basis for the internal currency, the volume of which by the rules of the gold standard must then be reduced. To accomplish this, the banking system must adopt a policy of deflation. The Central Bank discount rate will be raised, and by open-market operations the central bank will reduce the cash reserves of the commercial banks, which in their turn, in order to maintain their cash ratio, will reduce deposits by restricting their loans. As a result internal prices, including wages, should fall. This will make foreign goods relatively dearer, and so imports will be checked, while the export of home-produced goods will be stimulated because they will now be cheaper in world markets. In this way an adverse balance of payments will be rectified.

If, on the other hand, a country on the gold standard has a favourable balance of payments it will receive gold. This will increase the cash basis for the internal currency, and the banking system should then adopt a policy of credit expansion. As reflation occurs the demand for imports will increase and exports, now being dearer, will be discouraged. In this way a favourable balance of payments will tend to disappear. Figure 3.1 illustrates the self-regulating character of the gold standard, a deficit or surplus in the balance of payments being corrected 'automatically'.

The gold standard has provoked controversy from early times. The Mercantilist school, which flourished in the seventeenth and eighteenth centuries had one clear doctrine — that exports bring wealth to the nation. Under a gold standard, they asked what would keep one country from buying more goods from another country to the extent that they would eventually lose all their gold. The mercantilists wrote: 'A country will lose

DEFICIT	SURPLUS
Export of Gold	Import of Gold
Deflationary policy adopted	Reflationary policy adopted
Decrease in	Increase in
Income Prices	Incomes Prices
Decrease Increase	Increase Decrease
in imports in exports	in imports in exports
BALANCE	BALANCE

Figure 3.1. Balance of payments adjustments under the gold standard.

its gold unless the Prince introduces tariffs and quotas to cut down on imports of goods; and unless he gives subsidies to encourage exports.' David Hume in 1752, and economists ever since, have refuted this line of reasoning. First Hume noted that it could not be true that everyone would be losing gold under free trade. Where would it go — into the sea? And he demonstrated that it is no tragedy at all if a country goes permanently from having 10 million ounces of gold to having 5 million, over 1 million. If possessing half the amount of gold means merely that all prices are exactly halved, no one in the country is the least bit better or worse off. So, losing half or nine tenths of a nation's gold is nothing to worry about, Hume pointed out, if the nation merely ends up with an equivalently reduced price and cost level.

The Report of the Cunliffe Committee of 1918 (or, more exactly, the First Interim Report of the Committee on Currency and Foreign Exchange, of which Lord Cunliffe was chairman) tackled the specific British problem of restoring the gold standard after the end of the First World War, during which the gold standard was suspended. The Report was noted in particular for its support of the gold standard:

'The adoption of a currency not convertible at will into gold or other exportable gold is likely in practice to lead to over issue and so to destroy the measure of exchangeable value and cause a general rise in all prices and an adverse movement in the foreign exchanges.'

The Report provided what became a most influential account of how the gold standard was supposed to have operated before 1914. The following are the crucial passages:

'When the balance of trade was unfavourable and the exchanges were adverse, it became profitable to export gold. The would-be exporter bought his gold from the Bank of England and paid for it by a cheque on his account. The Bank obtained the gold from the Issue Department in exchange for notes taken out of its banking reserve, with the result that its liabilities to depositors and its banking reserve were reduced by an equal amount, and the ratio of reserve to liabilities consequently fell. If the process was repeated sufficiently often to reduce the ratio in a degree considered dangerous, the Bank raised its rate of discount. The raising of the discount rate had the immediate effect of retaining money here which would otherwise have been remitted abroad and of attracting remittances from abroad to take advantage of the higher rate, thus checking the out-flow of gold and even reversing the stream. . .

'But the raising of the Bank's discount rate and the steps taken to make it effective in the market necessarily led to a general rise of interest rates and a restriction of credit. New enterprises were therefore postponed and the demand for constructional materials and other capital goods was lessened. The consequent slackening of employment also diminished the demand for consumable goods, while holders of stocks of commodities, carried largely with borrowed money, being confronted with an increase of interest charges, if not with actual difficulty in renewing loans, and with the prospect of falling prices, tended to press their goods on a weak market. The result was a decline in general prices in the home market which, by checking imports and stimulating exports, corrected the adverse trade balance which was the primary cause of the difficulty.'

The reality of the Cunliffe view of gold standard adjustment is discussed further on p. 56 *et seq.* It is quoted in detail here as this view contributed to Britain's return to the gold standard in April 1925.

The Emergence of a World Wide Gold Standard before the First World War

The gold standard originally developed in the United Kingdom. The background of its evolution from a loose bimetallism goes back to the seventeenth and eighteenth centuries. Bimetallism is the monetary system in which two commodities, usually gold and silver act as legal tender to any amount at a fixed ratio to each other. Sir Isaac Newton, as Master of the Mint, established a fixed price of gold at £3.17s.10½d. per troy ounce, a price which remained fixed (with occasional breaks) for some 200 years. This standard, adopted somewhat unconsciously, was formally accepted in 1816. The Bank Restriction Act, in force from 1797 to 1821, forbade the Bank of England from redeeming its notes in gold. The Coinage Act of 1816 authorised the gold sovereign, a 20 shillings or £1 gold piece which was first coined the following year. Full redeemability in coin was achieved in 1821. The paper Pound was equivalent to the gold sovereign and England was on a full gold standard. Up until 1914 there were, in fact, no Bank of England notes for less than £5; a formidable sum in those days equivalent to rather more than three weeks earnings for an average man.

In the USA the Coinage Act of 1792 had defined a gold dollar and a silver dollar 15 times as heavy. (Gold dollars were not actually minted until 1849, but the gold content of the dollar was implied by the weight of gold coins of larger denominations.) The American mint ratio of 15 to 1 clashed with a ratio of about 15.5 to 1 prevailing on world markets. Since the US mint was accepting gold at less than its market value, little gold came to it for coinage, and the USA was in effect on a silver standard. The Coinage Acts of 1834 and 1837 reversed this disparity by slightly cutting the gold content of the dollar and setting a new mint ratio of very nearly 16 to 1. Since the USA was now valuing silver less highly relative to gold than was the world market, little silver was offered for coinage, and the USA was in effect on a gold standard. However, the unimportance of the USA in world trade combined with the effects of the American Civil War meant that the world was not yet on an international gold standard.

Table 3.1. Dates of Adoption of the Gold
Standard by Selected Countries.

Great Britain	1816
Germany	1871
Sweden, Norway and Denmark	1873
France, Belgium, Switzerland, Italy and Greece	1874
Holland	1875
Uruguay	1876
United States	1879
Austria	1892
Chile	1895
Japan	1897
Russia	1898
Dominican Republic	1901
Panama	1904
Mexico	1905

In the hope of promoting an international standardisation of currencies on a bimetallic basis, the Latin Monetary Union was formed in 1865 by countries whose standard monetary units were equal to the French franc: France, Belgium, Switzerland and Italy. Greece joined in 1868. Each member was to issue standard coins in denominations of 100, 50, 20, 10 and 5 francs in gold and 5 francs in silver and these coins were to circulate freely throughout the Union. In addition, each member country could mint subsidiary coins in amounts up to 6 francs per inhabitant, and public offices of all member countries were to accept these coins in payments of up to 100 francs.

The situation around 1870 was, in short, still far from being an international gold standard. The United Kingdom was fully on gold. Several legally bimetallic countries were in effect on gold. Germany, Holland, Scandinavia, and the Orient still had silver standards. Various wars and revolutions in the period 1848-1871 had inflated several important countries, including Russia, Austria-Hungary, Italy and the USA, onto irredeemable paper.

In the early 1870s movement toward an international gold standard gained momentum. As can be seen from Table 3.1 the gold standard was, in the period 1871 until 1905 adopted in most of the world's major trading countries. Thus the Europe of 1914 had ten currencies, all with fixed gold parities and fixed exchange rates. Various reasons for this increased adoption were discussed in Chapters 1 and 2. They include the finding of gold in the California and Klondike-Yukon areas and its subsequent flow into Europe in order to finance the American Civil War and the introduction of the cyanide process for refining gold. Both these developments substantially increased the supply of gold in the world.

The Gold Standard between the World Wars

The use of gold coins as money was virtually brought to an end by the First World War. In all belligerent countries gold coins were withdrawn from circulation and gold reserves were concentrated in the central banks. In contrast to the situation in 1914 the Europe of 1920 had twenty-seven paper currencies, none with a gold parity, none with fixed exchange rates and several of them in various stages of inflation. The First World War destroyed Britain's unchallenged financial dominance. Her already noticeable lag behind newly industrialised countries in production and export growth grew worse. Wartime trade interruptions spurred industrialisation in several of Britain's traditional markets. Britain lost ground in shipping and her role as an international investor was now much diminished. The American dollar destroyed the pre-eminence of sterling without itself taking over sterling's old role. The USA became an international financier during the First World War, lending abroad to finance an export surplus during its neutrality, and later to aid its Allies.

The dollar returned with ease to the full gold standard as early as 1919 and thereafter was the guidepost for realignment of other currencies. Rebuilding the gold standard was a gradual world-wide process. Some countries sooner or later regained their prewar gold parities; besides Great Britain, these included the British Dominions, Switzerland, The Netherlands and her colonies, Argentina, the three Scandinavian countries, and Japan (which did not drop her wartime gold export embargo until January 1930). By the end of 1925 some 35 currencies in addition to the US dollar had either been stabilised on gold or had displayed exchange rate stability for a full year. Three years later the apparent reconstruction of the international gold standard was substantially complete.

From 1920 until 1924 the sterling dollar exchange rate was left free to float. The wide fluctuations which followed combined with the German Mark being stabilised against gold in 1928 led to a widespread British preference for again fixing the exchange rate. In his budget speech of April 28th 1925 the Chancellor of the Exchequer, Winston Churchill, announced that the restrictions on gold and silver exports would lapse at once and that the Bank of England would redeem legal tender in gold for export. The Gold Standard Act of 1925, passed on May 13, required the Bank of England to sell gold ingots of not less than 400 fine ounces for legal tender at the traditional price of £3 17s. 10½d. per ounce 11/12 fine. Sterling could thus be redeemed in gold in amounts worth no less than almost $8300 (£1670). Redemption in coin by the general public was not permitted. Only the Bank of England was to have the right to bring gold bullion to the mint for coinage. The Bank remained obliged, as it had been without interruption ever since 1844, to buy all the gold offered to it at £3 17s. 9d. per ounce 11/12 fine. This system was known as the 'gold bullion standard'. It enabled all the advantages of the gold standard to be enjoyed without the expense of maintaining a gold coinage. Thus the essential features of the gold standard were retained: the standard unit of the currency still represented a certain weight of gold, and its price was fixed.

By 1926 only one of the world's major countries (Sweden) provided for the redemption of bank notes in gold coins; and only one (Great Britain) for the redemption in gold bullion. The widespread usage of gold coins had disappeared. The new gold-exchange standard, during the first years of operation in the 1920s, used chiefly the US dollar as a reserve asset; and for a few years, between 1927 and 1931, the pound sterling became the major reserve asset of the Netherlands.

The United Kingdom decision to return to the gold standard at the pre-war parity of \$4.86 (the actual parity was \$4.8665 but for convenience \$4.86 is used in the text as an approximation), announced in Churchill's budget speech, has since been interpreted as a watershed in British inter-war economic history. It was a decision which was widely criticised as having overvalued the sterling/dollar exchange rate. Most prominent among British critics was John Maynard Keynes. In 1925 Keynes published a pamphet entitled 'The Economic Consequences of Mr Churchill' warning that the old parity would hamper exports and necessitate further painful deflation in Britain. Churchill appears to have been impressed by four main arguments — the political risks involved in any decision not to return; the big, smooth rise in the exchange rate since the Tories' return (mentioned below); the widespread view that return would ensure long-term benefits on employment and trade, even if the short-term was hard; and the fact that after October 24, 1924, the gold supported exchange value of the pound had been able to rise, without apparently, any of the traditional deflationary side effects.

Although it is difficult to be precise about the extent of the over-valuation Moggridge has argued that the degree of overvaluation was 10 per cent, i.e., that British prices had fallen sufficiently only to make the pound worth some 90 per cent of this dollar rate.[1] The majority of accounts of this decision take the view that both the rate of exchange and the associated tight monetary policy imposed a burden upon the economy which, owing to structural defects, it was ill-equipped to bear and which made subsequent adjustment more difficult than it needed to be. Although they differ in the details of their reservations dissenters from this view are Alford,[2] Aldcroft,[3] Sayers[4] and Youngson,[5] with Moggridge[6] and Howson[7] being two of the most vocal critics of the exchange rate adopted. Sayers' view, that whatever the rate chosen was to a certain extent im-material, since this would be undercut by the French and Belgians, was seen to be correct when both the French and the Belgians returned to the gold standard in 1927 at an undervalued rate. An often forgotten point about the 1920s is that, after the initial inflationary burst following the end of the First World War in 1918, British economic policy was deliberately geared for a number of years to producing conditions for the restoration of the Gold Standard. This meant that by the end of 1922 deflationary policies had already nursed the exchange rate back up from $3.40 to $4.63, well before the formal revaluation to $4.86 on 3 April 1925.

The gold bullion standard lasted for six years, from 1925 until 1931. The overall inadequacy of the reserves for gold standard countries meant that countries with overvalued currencies were forced to pursue politically unacceptable deflationary policies. The rapid growth of trade unionism had made wages and prices much less flexible in the downward direction which further exacerbated unemployment. Governments generally were less willing to pursue 'laisser-faire' policies when lengthening dole queues threatened political stability. Great Britain abandoned convertibility in 1931 and the United States in 1933. In the United States during this depression period a considerable amount of gold coin and gold certificates went into hoards. After the dollar became *de facto* depreciated, Congress abrogated the gold clause in contracts and made all obligations payable in legal tender, that is, in any form of currency or coin. The Gold Reserve Act of 1934 established a new gold standard for the United States at $35 a fine ounce with gold used only for international settlements. Gold could no longer be coined and private holding of gold coin by US residents was forbidden, a prohibition abolished in 1974. (*This was discussed on pp. 40-41.*)

During the Great Depression (1929–1933) the international gold standard collapsed, and from 1933 to 1937 countries followed a policy of extreme nationalism. Currencies were devalued competitively, tariffs were raised and other restrictions on trade, such as exchange controls and multiple currency devices, were introduced. Currency devaluations intensified the depression in the gold-bloc countries i.e. France, the Netherlands, Belgium and Switzerland.

Currency devaluations began early in the Depression. In September 1931 sterling was depreciated by about 30 per cent. On January 31, 1934, the dollar was formally devalued by 41 per cent and, between 1935 and 1936, the currencies of the gold-bloc countries were devalued. The gold standard, so laboriously restored from 1925 to 1930, was abandoned by every great trading country by 1936. The great depression swept away not only the newly established parities but also the historic gold parity of the dollar, sterling, the Netherlands guilder, and the Swiss franc. It was not until the link between gold and currencies was severed that the depression could be brought to an end and a slow recovery could begin.

The inter-war years can be seen, with hindsight, as the period when the USA achieved economic dominance over the Western world. The failure of international monetary relations during these two decades was probably due to the USA's reluctance to assume the responsibilities that this position forced upon it. By the end of the Second World War, it had overcome these inhibitions. The formulation of a new spirit of international monetary co-operation, a necessary condition for the planning of the post-war reconstruction, depended, to a certain degree, upon the willingness of the USA to assume economic leadership.

The Gold Standard after the Second World War

Just as opinion at the end of the First World War was almost unanimously in favour of a return to the gold standard, so there was virtual unanimity at the end of the Second World War against returning to a gold standard without substantial modifications. There was also a general consensus on the need to ensure the convertibility of currencies between countries and to avoid large and frequent fluctuations in exchange rates. The Bretton Woods agreement (1944) was an attempt to modify the gold standard with the intention of removing the supposed defects of the classical gold standard while retaining some of its better features. Appendix I summarises the similarities and differences between the British and American plans which were put forward at Bretton Woods. The system adopted at Bretton Woods became known as the gold exchange standard.

According to its Articles of Agreement, the purposes of the International Monetary Fund (IMF), which was established at Bretton Woods, were to promote international monetary co-operation, facilitate the expansion of international trade for the sake of high levels of employment and real income, promote exchange-rate stability and avoid competitive depreciation, work for a unilateral system of current international payments and for elimination of exchange controls over current transactions, create confidence among member nations and give them the opportunity to correct balance of payments maladjustments while avoiding measures destructive of national and international prosperity.

Under the provisions of the Bretton Woods Agreement, all countries were to fix the value of their currencies in terms of gold but were not required to exchange their currencies for gold. Only the dollar remained convertible into gold at $35 per ounce or 0.0285714 fine ounces of gold per dollar, which is the same thing. Therefore all countries decided what they wished their exchange rates to be *vis-à-vis* the dollar, then calculated what the gold par value of their currencies should be to give the desired dollar exchange rate. All participating countries agreed to try to maintain the value of their currencies within 1 per cent of par by buying or selling foreign exchange or gold as needed. The rules of the IMF allowed members' monetary authorities to sell gold (but not to buy) above the official price and to buy gold (but not to sell) below it. This system engendered the exchange rate stability provided by the gold standard.

Members were required to pay 25 per cent of their IMF quota in gold (with concessions for countries holding small gold stocks). It was envisaged that gold would continue to form the major part of international reserves and that gold transactions would play an important role in settling those international payments that could not be offset against one another through exchanging one convertible currency for another in the foreign exchange market.

In all these respects the new system was similar to the old gold standard. However, the IMF agreement provided formal rules allowing members to change the par values of their currencies to meet a 'fundamental disequilibrium' in their balance of payments, though these words were not defined. The Fund provided members with the opportunity of 'drawing' (i.e. borrowing) foreign currencies in amounts related to their quotas to supplement their reserves, thus allowing more time for the correction of balance-of-payments deficits; and the 'scarce currency clauses' (which never came into operation) were intended to put pressure on countries with a surplus in their balance of payments to contribute their share towards adjustment.

The most important difference between the new system and the old was not written into the Bretton Woods Agreement at all but followed from the policies of individual countries. No country, after 1945, operated any effective link between its gold reserve position and its domestic money supply; where a formal rule was maintained it was completely ineffective. The USA retained a formal link in the form of a 25 per cent reserve ratio, but in the early post-war years the US gold stock was so large that the rule imposed no great constraint; when the fall in the gold stock and the expansion of the money supply produced a situation where the rule might soon have begun to 'bite', Congress repealed it, in 1968. In the United Kingdom the Bank of England still operates within the legal form of the 'fixed fiduciary issue'; but the gold stock is now held by the Exchange Equalisation Account, not the Bank, and the amount of the note issue is in practice varied at the discretion of the Treasury. Thus fluctuations in reserves no longer have the effect on the domestic money stock that an inflow or outflow of gold had under the old system, and adjustments of the balance of payments depend on discretionary action by governments. This was the important difference between the new and the old systems.

The transitional period after the setting up of the IMF proved much longer than was envisaged at Bretton Woods and it was not until the late 1950s that the new system was fully operational. After being closed, or all but closed, since 1939, the London gold market reopened in March 1954. This was a market in which private buyers and sellers could operate. The market was calm for 6½ years and the gold price remained within a few cents of the US Treasury's buying and selling prices of $34.9125 and $35.0875 per ounce. In 1960 following the Cuban missile crisis there was a large rise in the speculative demand for gold prompted by rumours of a de-valuation of the US dollar (equivalent to a rise in the official gold price). On October 20 1960 the London price reached $40 an ounce, despite large official sales, mostly by the US Treasury. This rise in the gold price caused foreign central banks, fearing a dollar devaluation, to cash in surplus dollars at the United States gold window, and the US Treasury found itself forced to put out a press statement formally endorsing Bank of England sales of gold to stabilise the London price at the ultimate expense of the American gold stock. By the autumn of 1960 the weekly Treasury reports of American losses through the gold window were making the front page of the *New York Times.*

Charles Coombs, formerly Senior Vice President of the Federal Reserve Bank of New York who was responsible for US Treasury and Federal Reserve operations in the gold and foreign exchange markets summarised the situation as follows:[8]

'Thus in late 1960 the dollar had become convertible into gold on demand not only by central banks but also by private speculators all over the world. The United States Government had become thoroughly trapped by its gold commitments under Bretton Woods and a heavy responsibility for devising a safe escape route now fell on the new team of Treasury officials to be appointed by President-elect John Kennedy.

In February 1961, Kennedy made his famous pledge that the dollar-gold price would remain immutable. The price settled back to $35 but the pledge was no more than a holding operation and negotiations be-tween the leading Central Banks immediately began which were to result in the creation of the Gold Pool. This was an agreement whereby the Bank of England would sell gold on behalf of the central banks of the USA, the United Kingdom, Switzerland, Belgium, France, Germany, Italy and the Netherlands, in sufficient quantities to keep the price at $35. The Bank of England bought or sold in the market, as required, to stabilise the price and the participants contributed gold for sale, or shared gold purchased, according to a quota. France dropped out in 1967 but the remaining countries operated the pool until March 1968.

Between 1961 and 1965 there were substantial Russian sales through the pool thereby helping the pool to operate both ways with purchases somewhat exceeding sales over the period as a whole. After the end of 1965, Russian sales ceased and the pool became a substantial net seller. The climax came with the devaluation of sterling in November 1967; this again cast doubt on the stability of the dollar, which was widely regarded as overvalued, and created an enormous speculative demand for gold. The question now was how much official gold was to be diverted to the London gold market before it was acknowledged that the game was up. Several attempts were made to dam the flood, but with gold market speculation now focussing squarely on the dollar the crisis came in March

1968. On Monday March 11th, gold pool losses were $118 million, on Tuesday $103 million, Wednesday $179 million, and by lunchtime Thursday $220 million. At 3.00 pm on Thursday the 14th March President Johnson discontinued pool operations and it was agreed that the London gold market should remain closed for two weeks to allow time for a solution to be found. The cost to the gold pool participants of the last six months of its operation had been some $2.8 billion.

As Coombs again points out the time differential between New York and London led to more than the usual complications involving decision taking which significantly affects the interaction of monetary relations.[9] On the occasion of the closing of the gold pool there was the formal necessity of waking up Queen Elizabeth around midnight London time to get her permission to close the gold market. In addition there was the decision of British Prime Minister Harold Wilson not to wake up his Foreign Minister, George Brown, with the result that the latter resigned in protest the next day.

The Governors of the gold pool banks met in Washington on Saturday 16th March and settled for the long-canvassed proposal for a two-tier system. The Central Banks agreed to buy and sell gold among themselves at $35 but not to deal in the London market which would be left free to find its own level.

As the London gold market reopened on April 1 the price settled around the $38 level in quiet and orderly trading and for the rest of 1968 it did not rise above $43. In 1969, moreover, a renewed flow of gold from South Africa combined with the overhang of private hoards to bring the price down to the official $35 parity. Mainly reflecting European pressures, an IMF agreement was then devised to effectively place a floor under the London gold price.

When President Nixon, looking to re-election in 1972, wanted to expand demand to pull America out of the minor world recession in 1970-71, speculation against the dollar mounted. This forced central banks in continental Europe and Japan to buy huge quantities of dollars rather than see their exchange rates rise (which would have hit their exporters and so intensified their recessions). The free market gold price rose sharply. The effects of this upon the international monetary system were intolerable. The greater the gap between the two prices, the official price and the free market price, the more speculation increased, based upon the assumption that there would eventually be a revaluation of the official price. The dollar, at least, in principle, was convertible at the official price so the outflow of US gold reserves accelerated. All newly mined gold production was going to the free market, hence there was none available to increase international liquidity. The disparity between the two prices prevented the central banks from utilising their reserves as a means of international settlement. They had no wish to ship gold at what was so obviously an unrealistically low price. The USA, with $10 billion in gold reserves against debts of $50 billion in other countries' reserves, decided to suspend convertibility in August, 1971. The gold link was broken and the dollar set to float. See Figure 3.2 for details of the US gold reserves and the US external liabilities.

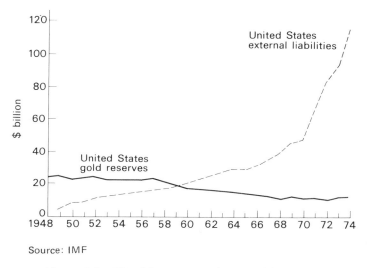

Source: IMF

Figure 3.2. US gold reserves and external liabilities.

Despite revaluations of the official price to $38 per ounce (December 1971) and $42.22 per ounce (February 1973), the market price was consistently above the official price. As already mentioned, in these circumstances central banks were not prepared to use gold for settlement purposes and gold stocks became effectively immobilised.

Official Gold Market Developments in the 1970s

Gold's role in the world monetary system was formally reduced when, at the Jamaican Conference of January 1976, it was decided to abolish the then prevailing fiction of an official gold price. The major central banks undertook not to increase their aggregate gold stocks for two years from February 1976, and it was decided that, through a four-year programme of auctions, one-sixth of the IMF's gold stock was to be returned to its members, and a further one-sixth sold off, with the profits, (the difference between the market price and $42.22 per ounce), going to developing countries.

This decision was formally set out in the April 1978 Second Amendment to the IMF's Articles of Agreement. The Amendment makes clear the fund's intent to reduce the importance of the role of gold in the fund, and to enhance the role of the Special Drawing Right (discussed in detail in Chapter 4) for official transactions, in anticipation of the SDR eventually becoming the principal reserve asset of the international monetary system.

The Second Amendment states that members may no longer define the exchange value of their currency in terms of gold. The official price of gold (SDR $35 per fine ounce) was formally abolished, and member nations were set free to trade in gold at any price consistent with their domestic legislation. Members were no longer required to pay gold to the fund in connection with any transaction or operation, and the fund was to accept gold from a member only by a decision requiring a majority of 85 per cent of the total voting power, a percentage high enough to give the USA in effect a veto power. Members with accrued obligations to pay gold to the fund at the date of the Second Amendment were required to discharge such obligations with SDRs, or in the currencies of other members prescribed by the fund. Official management of the price of gold was forbidden.

However, the extreme international financial imbalances of the last decade have rendered impractical any hope that countries owning gold would not take advantage of this fact to meet balance of payments deficits. Indeed, the IMF (via the Committee of Twenty) affirmed as early as 1975 that gold remained an important part of national reserves, and should be used to finance deficits when necessary. The Zeist and Martinique Agreements of 1974, in fact, permitted gold to be pledged, and accepted as collateral by central banks, at something related to market prices.

The evolution of gold's monetary role took a new direction with the inauguration of the European Monetary System (EMS) in March 1979. (*For an explanation of the EMS see Appendix II.*) All participants in the EMS are required to deposit with the European Monetary Co-operation Fund, which since April 1973 has acted as a clearing house for central banks' claims and debts stemming from intervention within the EMS, 20 per cent of their dollar reserves plus 20 per cent of their gold reserves. In return, participants receive claims on the EMS, denominated in ECUs (European Currency Units). The transfer takes place by three-month swap agreements between central banks, automatically renewable during the transitional period of two years following establishment of the EMS. The key point, here, is that the ECU, which may eventually evolve into a new international reserve asset, is backed in part by gold. (Developments in the gold price during the 1970s are discussed in Chapter 5.)

A Return to the Pre-First World War Gold Standard?

A significant body of opinion wishes to see a return to the classical gold standard. The gold lobby has one of its most eloquent spokesmen in

William Rees Mogg, the then Editor of the London *Times*. In a review of R. W. Jastram's book, *The Golden Constant*, in *The Times* of 14th July 1978, Mogg concludes thus:

> 'The paper systems seem now to be coming to the end of their effective lives; the restoration of gold to a central position in the world money is the most valuable reform that could be undertaken.
>
> 'The historic evidence is that gold is greatly more efficient than paper in producing stable prices and stable and low interest rates. These are the optimum monetary conditions for economic development and social stability. Gold has in the past provided a guarantee for centuries against world inflation; it would do so again.'

Thus a system of fixed exchange rates, with national governments committed to their maintenance, and gold as the sole reserve asset would, it is argued, permit a return to monetary discipline. Appeals for a return in the 1980s to the gold standard serve as a convenient slogan, or rallying point, for opposition to the twentieth century record of inflation. A return to the gold standard is sometimes seen as a means of stopping the state, by the mechanism of monetary inflation, from wresting resources from its subjects and building up its power at their expense. Advocates of a return, however, do tend to be somewhat vague about what precisely the mechanisms of the new system would be and in addition tend to have a somewhat idealised view of the nature of the gold standard as it actually operated before the First World War. It is in consequence necessary to distinguish between the myths and the reality of this gold standard in order to discover what can be learnt for re-introducing the gold standard as the basis of the international monetary system.

Proponents of a return to the gold standard are somewhat vague as to precisely what they understand by it. This is one reason why Milton Friedman, a Nobel prizewinner in Economics, has insisted on the distinction between 'real and pseudo gold standards'. As he has pointed out,[10] from 1879 to 1933 in the USA, gold circulated as money, and paper money was redeemable in gold.

> 'This made for a close link between money and gold. During these 54 years, the amount of money was never more than twelve times that of gold, never less than five times. Withdrawal of gold from circulation and prohibition of the private ownership of gold in 1933 combined with developments in the rest of the world to convert gold from the kingpin of the monetary system into a commodity whose price is pegged. After President Roosevelt raised the official price of gold from $20.67 to $35 an ounce in 1934, our gold stocks grew explosively until the amount of money equalled only two and half times the amount of gold. Then, as other prices rose, $35 an ounce became cheap and our gold stocks declined. However, the quantity of money continued to rise

until it is now 30 times the amount of gold. The link between gold and the quantity of money has become a rubber band. A real gold standard has become a pseudo gold standard.'[10]

A small minority of gold standard advocates (sometimes referred to as 'gold bugs') want a 100 per cent gold standard. Gold would be the only money. Government issue of any other kind would be forbidden, and even private banknotes, demand deposits (current accounts in the United Kingdom), or other liabilities usable as media of exchange would be banned unless fully backed by gold. Less rigorous proposals would forbid government fiduciary money (i.e. notes unbacked by precious metals) but tolerate the private creation of fractional reserve money. Some proposals would impose reserve requirements on banks; others would subject them to no regulations. The gold reserve ratios that banks would voluntarily adopt to be safe and to guard their reputations for soundness would, it is hoped, gear the total supply of money, in a stable enough manner, to the unmanipulated total stock of gold. Still other proposals would tolerate fiduciary issues not only by commercial banks but also by governments or central banks.

Many myths about the pre-First World War gold standard exist. It is a myth that the monetary use of gold increased during the gold standard era (1880-1914). Indeed, the monetary use of gold within countries declined sharply throughout the nineteenth century and it accounted for less than 10 per cent of domestic money supplies by 1913. By 1913 in fact foreign exchange represented about 20 per cent of world reserves. It is also a myth that exchange rate changes were unknown in this period. Some countries around the periphery of the system (including Spain) did not peg their exchange rates at all. Devaluations were frequent, especially in Latin America. It is a myth that countries let external pressures determine their internal policies. Downward wage adjustments were extremely small in the few cases in which they did occur, and it appears that domestic economic policy changed in response to the requirements of external balance less than half the time. Bloomfield found such 'neutralisation' present in 60 per cent of his total observations for this period.[11] However, it is true that no major country changed its exchange rate during this period. Bergsten puts forward four reasons as to the combination of events which made this adjustment process possible.

The first explanation is that official reserves grew steadily and quite rapidly at least from 1880 through 1913. The data are far from adequate, but the best available evidence suggests that world holdings of gold and foreign exchange reserves grew by about 7 per cent annually during this period, from about $1.2 billion in 1880 to about $6.5 billion in 1913, and by almost 10 per cent annually during its last decade (*see Table 3.2*). The increases appear to have been distributed fairly widely among countries. Indeed only Denmark, of ten countries surveyed by Bloomfield, experienced a decline in the ratio of its reserves to imports over the period as a whole, and several (including the US) experienced sizeable increases. The gross reserves of each of the three key currency countries – Britain, France and Germany – rose steadily. Thus, pressure for rapid adjustment, which would prompt the application of controls but also force exchange rate changes, was significantly cushioned by the growth of owned reserves.

Table 3.2. World Reserves Prior to World War I
(in billions of dollars)

	1880	1885	1903	1910	1913
Gold	1.0	1.3	2.6	4.2	4.9
Foreign Exchange	0.2	0.3	0.8	1.5	1.6
Total	1.2	1.6	3.4	5.7	6.5

Sources:
Arthur J. Bloomfield, *Short-term Capital Movements under the Pre-1914 Gold Standard*, Princeton Studies in International Finance, No. 11, July 1963.
Robert Triffin, *The Evolution of the IMF: Historical Reappraisal and Future Objectives*, Princeton Studies in International Finance, No. 12, June 1963.
Peter H. Lindert, *Key Currencies and Gold: 1900–1913*, Princeton Studies in International Finance, No. 24, August 1969.

A second explanation is that a major source of assistance was provided by the tendency of the four main trading countries of the world, France, Germany, USA and Britain, collectively accounting for half of world trade, to follow much the same course of the business cycle. All four were in the same phase of the cycle in over half of the whole period between 1879 and 1914, while the three European countries were in the same phase in over four-fifths of the period. For much of the time, moreover, the movement of economic activity and of interest rates were in the same direction in most countries, as activity changes spread from these four main countries. As Morgenstern has shown, the three major powers in the system, Britain, Germany and France, were together in the expansion or contraction phase of the business cycle an amazingly high 83 per cent of the time from 1880 through mid-1914.[13] This parallelism tended to ease the problems of balance of payments adjustment. As one country experienced rising incomes and imports and a worsening balance of payments, the parallel expansion of incomes and import purchases in other countries brought the initial country rising exports and a mitigation of its balance of payments difficulties. Thus the adjustment mechanisms working through prices and incomes had a powerful effect in triggering adjustment. The dilemma cases of rising domestic unemployment and a worsening balance of payments which may have necessitated exchange rate adjustments were therefore infrequent. Indeed, as Bergsten points out, due to the lack of official statistics in those times the potential conflict among the different objectives of national economic policy was reduced by the fact that individuals in one country could not observe much better unemployment or price performances in other countries.[14] With the rapid growth of official statistics and their widespread publicity this latter situation has substantially altered.

Thirdly and most importantly, national authorities had not yet become responsible for preserving domestic full employment or even price stability. Hence there were few efforts to offset the adjustment impulses of the highly synchronised international business cycle. Indeed, most central banks were privately owned and they worried more about their own liquidity and earnings than about public policy. Fiscal policy barely existed. The simple target of preserving a fixed exchange rate dominated policy and was not 'undermined' by active manipulation of other policy instruments. The traumatic experiences of the twentieth century with inflation and mass unemployment, which both underlie the different national policy preferences that exist today compelling governments to take a far more active economic policy role, were still to come.

Finally there was a bias against controls over external transactions during this time period thus tending to place the emphasis for adjustment on changes in the internal economy rather than changes on the external economy control mechanisms (*viz*. exchange rate changes and exchange control regulations). The biases against controls were reinforced by the intellectual ascendancy of laissez-faire economics, and by the successful application of those ideas by the United Kingdom. In the late nineteenth century London was established as a dominant centre for international finance although by 1900 France and Germany were both major currency centres. During this gold standard period international reserve assets were sometimes said to have consisted of gold alone, but actually they consisted of gold and sterling deposits in British banks. Most of world trade, estimated at more than 90 per cent by Whitaker, was financed in London.[15] The world's short-term and long-term capital markets were located in London, together with the gold market and produce markets, while the bill on London amounted to world currency. This concentration reinforced the role of Britain in world trading relations, and facilitated the spread of the British monetary area to cover much of the British Empire. Simultaneously there was the expansion of banks which kept considerable balances in London (the start of the sterling balances) providing both their cash reserves and the foreign exchange reserves of the territories in which they operated. Hence, deficits incurred over a considerable range of Britain's international transactions meant that gold would never be taken, merely that such 'London' balances would rise. Other international bankers, too, would in the short run be quite prepared to see their London balances rise if it was made worthwhile. Such arrangements, together with the great confidence in sterling (it was good as gold, or even better, since interest could be earned on it), make it clear that underlying the pre-1914 gold facade was a considerable sterling support structure.

Britain's commercial and financial leadership stemmed in turn from a combination of historical factors, including the policy of free trade, which was firmly established by the middle of the nineteenth century; and a head start in industrialisation, which generated wealth and savings available for loan and investment abroad.

There is reason to doubt the supposed automaticity of the gold standard particularly when allowance is made for the fact that it was in reality a sterling standard. Arthur Bloomfield demonstrated that the national governments were not as passive in their domestic economic policies as tradition would suggest.[16] The so-called 'rules of the game' were often broken, as governments sought to neutralise the effects of the adjustment mechanism. Bloomfield illustrates that central banks indulged in open market operations to cancel out any effects of both gold outflows and inflows. That is to say when an export of gold reduced the liquidity base of an economy, instead of allowing this to drive up interest rates as the classical system would dictate, the central bank would neutralise the influence of this outflow by buying gilt-edged securities from the commercial banks, thereby maintaining the level of the money supply.

Thus, although the immediate impact of a balance of payments deficit is to reduce the quantity of money in circulation (since the private sector will purchase foreign currency with domestic currency in order to pay for the deficit) this impact may be offset by purchase of government bonds by the monetary authorities thereby pumping money back into the economy. This policy, known as sterilisation of the financial effects of the balance of payments was fairly standard in the 1880–1913 time period and was also frequently undertaken in the post-First World War time period. The adoption of sterilisation policies does not eliminate the drain of private sector wealth. But these policies do ensure that this drain is in the form of non-monetary wealth so that any decline in private sector expenditures, and therefore the corresponding balance of payments adjustment mechanism, is likely to be correspondingly slower and weaker.

Bloomfield further questioned the role of private short-term capital movements. Unquestionably, short term interest rate differentials and exchange rate fluctuations within the gold points played a dominant role in directing the flow of private short term funds between gold standard countries, even if the degree of mobility of the funds so motivated has sometimes been exaggerated. But by no means all private short-term capital movements could be explained in these simple terms. Preferences, based on institutional arrangements or long standing banking connections, as to foreign lenders or as to markets in which to place short-term funds; the availability of credit as contrasted with its cost; the requirements of external debt service; the changing needs for maintaining working balances in given centres; consideration of bank liquidity – all these and other factors exerted an influence on the volume and direction of short-term capital flows that may at times have overshadowed interest-rate factors.

Conditions almost unique in history smoothed the operation of the system. Money, wages and prices were probably more flexible in both directions than they later became. Relative calm in social and political affairs and the absence of ambitious programmes of government spending and taxation all tended to tighten monetary control. Moreover, as Triffin has made clear, credit money, i.e. bank currency and deposits, played an overwhelming part in sustaining feasible rates of economic growth in the century before the First World War.[17] Triffin estimates that by 1913 credit money accounted for about 85 per cent of the world money supply.

Heller and Kreinen showed that balance of payments adjustments via domestic macro-economic policy (*à la* gold standard rules) for the 1960s time period would be costlier than adjustment via the exchange rate for most countries.[18] They conclude that the ratio of costs from income adjustment to costs from exchange rate adjustment is roughly 5:1 for Canada, 3.5:1 for Japan, and 3:1 for the United States. A similar study for the post OPEC world would undoubtedly indicate that this ratio has worsened. Michaely undertook a comprehensive study of the domestic policy responses to payments imbalances of the world's major countries during 1950–66.[19] This is in fact a test of whether countries would be willing to live by gold standard adjustment rules as the period under consideration was one of fixed exchange rates. He discovered that no country ever moved its fiscal policy consistently even in the direction called for by external considerations. Michaely also discovered that even the directional response of monetary policy to external considerations weakened from the 1950s to the 1960s.

As stated by Bergsten reasons for this antipathy to gold standard rules are obvious.[20] The level of domestic unemployment or inflation which would be needed to provide the sole cure for payments deficits and surpluses, respectively, would create huge economic costs and thus be politically unacceptable in most modern societies. Governments and monetary authorities already have great difficulty in achieving their numerous policy targets, and they could hardly afford to give up three of their policy instruments – the exchange rate, selective controls and external financing – which would be needed if the 'rules' of the gold standard were to be adhered to.

References

(1) Moggridge, D. E. *British Monetary Policy 1924–31* (Cambridge 1972).

(2) Alford, B. W. E. *Depression and Recovery in British Economic Growth 1918–1939.* (Macmillan 1972).

(3) Aldcroft, D. H. Economic Growth in Britain in the Inter-War Years: A Reassessment (*Economic History Review XX*. 1967).

(4) Sayers, R. S. 'The Return to Gold, 1925' in L. S. Pressnell (ed.) *Studies in the Industrial Revolution*. 1960.

(5) Youngson, A. J. *Britains Economic Growth* 1920–1966. (1967).

(6) Op. cit.

(7) Howson, S. *Domestic Monetary Management in Britain 1919–1938* (Cambridge University Press 1975).

(8) Coombs, C. A. *The Arena of International Finance* (John Wiley 1976).

(9) Op. cit.

(10) Friedman, M. *Dollars and Deficits: Inflation, Monetary Policy and the Balance of Payments.* (Prentice Hall. 1968).

(11) Bloomfield, A. J. Short-term Capital Movements under the pre-1914 Gold Standard, *Princeton Studies in International Finance, No. 11.* July 1963.

(12) Bergsten, C. F. *The Dilemmas of the Dollar.* University Press 1975. New York.

(13) Morgenstern, O. *International Financial Transactions and Business Cycles.* Princeton University Press 1959.

(14) Op. cit.

(15) Whittaker, A. C. *Foreign Exchange.* Second Edition (New York: Appleton 1933).

(16) Bloomfield, A., *Monetary Policy Under the International Gold Standard 1880–1913.*

(17) Triffin, R. The Evolution of the IMF: Historical Reappraisal and Future Objectives. Princeton *Studies in International Finance No. 24.* August 1969.

(18) Heller, R. H. and Kreinen, M. E. 'Adjustment Costs, Currency Areas and Reserves' in Willy Sellerkaerts ed. *Essays in Honour of Jon Tinbergen* (Macmillan: London 1972).

(19) Michaely, M. *The Responsiveness of Demand Policies to Balance of Payments: Post War Patterns* (Columbia University Press, New York 1971.)

(20) Op. cit.

APPENDIX I

POST SECOND WORLD WAR PLANS FOR REFORMING THE INTERNATIONAL MONETARY SYSTEM: THE ALTERNATIVE ROLES GIVEN TO GOLD

The lessons of the First World War were well remembered in the formation of international financial policy during and after the Second World War, and the monetary problems of the post-war period were not left to improvisation. Instead, the USA and the United Kingdom prepared independent plans for international monetary cooperation. These plans are best known under the names of their principal authors, the White Plan and the Keynes Plan. Other plans were proposed by Canada and France.

The Keynes Plan proposed the creation of an International Clearing Union (ICU) based on an international unit, the 'bancor', with a value fixed (but not unalterably) in terms of gold. The 'bancor' was to be accepted as the equivalent of gold by member countries in the settlement of international balances. Central banks of member countries would keep accounts with the ICU. The ICU would not have subscribed capital. Instead, countries were to be given overdraft facilities up to a prescribed quota based on the countries share of international trade. Countries with a surplus in international payments would accumulate credit balances; those with a deficit would build up debit balances. Moderate interest charges would be levied on both debit and credit balances. As the ICU would not be a closed payments system, with bancor holdings not convertible, the debits and credits would necessarily be equal and there could be no question of the adequacy of its resources for meeting the reserve credit requirements of its members, within the quota limits. The members of the ICU would agree among themselves on the initial par values of their currencies in terms of bancor. These could be changed thereafter only with the permission of the ICU. Under certain conditions of persistent deficit, the ICU could request a member to devalue its currency and could require it to repay part of its indebtedness out of its gold and foreign exchange reserves. In general, the ICU was not to concern itself with the domestic economic policies of its members.

The White Plan proposed the creation of an International Stabilisation Fund (ISF) to maintain orderly exchange arrangements. Each member country would have to agree on a par value of its currency defined in terms of an international monetary unit (unitas) equivalent to $10 in gold. Changes in parities could be made only after consultation and with the approval of the ISF. Members would not be permitted to impose exchange controls without the approval of the ISF and would be obliged to remove existing exchange controls. To provide the ISF with resources, members would be assigned quotas and would subscribe gold and their own currencies up to the amount of these quotas. Members could purchase foreign exchange from the ISF with their own currencies, up to one-quarter of their quotas annually, with maximum net credits not to exceed 100 per cent of their quotas. Drawings in larger amounts could be made only with special approval after waiver of the quota limits. Members would have to repay drawings in gold and convertible foreign exchange when their reserves increased.

The ISF would have been given wide powers under the White Plan. Members acquiring currencies in settlement of international payments could sell them to the ISF either for their own currencies or for foreign exchange. Thus, the ISF would have assured the convertibility of any currency acquired by a member in settlement of a balance of payments surplus. As the reserve credits of the ISF were to be given in specific currencies, the operations contemplated could exhaust its holdings of the currency of a country with a large and persistent surplus. The Plan, therefore, provided that if it appeared that a currency would become scarce the ISF would issue a report to the surplus country with recommendations designed to restore the ISF's holdings of the scarce currency.

The principal differences between the two plans may be summarised as follows. The ISF would have had wide powers of intervention through international financial operations undertaken on its own initiative. The ICU would have been a passive institution, providing reserve credit facilities only on the initiative of its members. The ISF contemplated the use of the foreign exchange market for settling international balances; the ICU contemplated the clearing of such balances through its accounts. The ICU would not itself have held gold or currencies, and the role of gold in international payments would have been circumscribed. The ISF would have had its own reserves of gold and currencies subscribed by members, and gold was to have a significant role as reserves and in international settlements. Finally, under the ICU the responsibility for restoring international equilibrium would be shared by debtor and creditor countries, while under the ISF it would have fallen primarily on debtor countries. The reserve credit facilities would have been far larger under the ICU and would have been available on more generous terms.

After two years of preliminary discussion, President Roosevelt invited forty-four countries to the Monetary and Financial conference at Bretton Woods, New Hampshire, which met July 1–22, 1944. The IMF followed in form the proposal of the USA, but included many details from the British and Canadian proposals.

APPENDIX II

THE EUROPEAN MONETARY SYSTEM

In March 1979, nine major European countries, members of the European Economic Community, launched a new experiment in international monetary cooperation – the European Monetary system. Its objective is to stabilise exchange rates between the currencies of the member states of the European Communities and to contribute to the strengthening of international monetary relations. At the same time it is intended to give a new impetus to the process of European integration. The EMS supersedes the European narrower margins arrangement ('snake') of 1972, and combined old and new rules for monetary relations in the European Communities. Seven of the nine European Community members, Belgium, Denmark, France, Germany, Ireland, Luxembourg and the Netherlands, became full participants. Italy decided to participate under modified conditions, and the United Kingdom, while becoming a member of the EMS elected not to participate in all the arrangements.

The three major aspects of EMS are the application of the parity grid as in the old 'snake', the role of the European Currency Unit (ECU) and the supporting credit facilities needed for intervention purposes.

The parity grid system

Par values for individual currencies are established for individual currencies and permitted fluctuations of ±2¼% around these are permitted for all participants except for Italy where the permitted fluctuation is ±6%. The exact rates are given in Table 3.3.

Thus these rates establish the moments at which central banks intervene in the foreign exchange market in order to keep the rates fixed. These cross rates are so established that, for example, when the Deutschemark is at its weakest point against the Belgian franc in Germany the Belgian franc is at its strongest point against the Deutschemark in Belgium. Continuous central bank intervention maintains the cross rates between the participating currencies.

The European Currency Unit (ECU).

A newly created monetary unit, the ECU is the linchpin of the new system. The ECU does not exist in the physical sense that currencies of individual countries do. It does serve, however, as a monetary asset that participating central banks can hold as reserves. The central banks can also loan and borrow the unit, and it can be used in settling debts between them. Though use of the unit is limited initially to countries participating in the EMS, it is expected that the ECU could serve eventually as an international reserve asset similar to the Special Drawing Rights issued by the International Monetary Fund and held and used by central banks worldwide.

In addition to its monetary function, the ECU will serve an accounting function, its value providing a benchmark against which the central rates of individual currencies of the EMS members will be established. Thus, at the inception of the EMS, each of the participating countries formally defined the value of its currency in terms of the number of units of that currency one ECU would 'buy'.

Valuation of individual currencies in terms of the ECU serves two purposes: (1) it establishes a 'central rate' for every currency in terms of other currencies, these relative rates forming a 'bilateral grid' of exchange

Table 3.3. Intervention Rates (November 1979)

	Deutsche Mark	French Franc	Dutch Guilder	Belgian Franc	Italian Lira	Danish Kroner	Irish Punt
Deutsche Mark	—	2.3033 2.4093	1.080775 1.1305	15.6740 16.3955	439.312 495.287	3.0423 3.1826	0.26323 0.27553
French Franc 10	4.1505 4.3415	—	4.5880 4.7990	66.5375 69.600	1,864.9 2,102.5	12.9150 13.5095	1.11739 1.16881
Dutch Guilder	0.88455 0.92525	2.0838 2.1796	—	14.1800 14.8325	397.434 448.074	2.75254 2.8790	0.23813 0.249089
Belgian Franc 100	6.0990 6.3800	14.3680 15.0290	6.7420 7.0520	—	2,740.44 3,089.61	18.9785 19.8520	1.64198 1.71755
Italian Lire 1,000	2.019 2.276	4.7560 5.3620	2.23175 2.5160	32.365 36.490	—	6.2825 7.0830	0.54354 0.61080
Danish Kroner 10	3.142 3.287	7.4020 7.7430	3.4735 3.6330	50.375 52.690	1,411.83 1,591.72	—	0.84592 0.84854
Irish Punt	3.6320 3.7990	8.5555 8.9495	4.0145 4.1995	58.2225 60.9020	1,631.85 1,839.78	11.3013 11.8214	—

rates linking all EMS currencies; (2) it provides reference points for establishing a 'threshold of divergence' that, once reached, will create a presumption for members to take specific economic measures.

In purely technical terms, the ECU is a composite unit consisting of the European Community member currencies. Its composition is given in Table 3.4.

Table 3.4. Currency Weights of the ECU

	In Absolute Amounts of Currency	As Percentage Shares on the Basis of Market Rates on March 1, 1979
Deutsche mark	0.82800	33.02
French franc	1.15000	19.89
Pound sterling	0.08850	13.25
Netherlands guilder	0.28600	10.56
Italian lira	109.00000	9.58
Belgian franc	3.66000	9.23
Danish krone	0.21700	3.10
Irish pound	0.00759	1.11
Luxembourg franc	0.14000	0.35

The weights assigned to each currency in the basket are derived from the relative GNP of each member country and that country's share in intra-European trade. The weights will be reexamined every five years, or if the relative value of any currency changes by 25 per cent, the weights will be reexamined on request.

The threshold of divergence

The exchange rate system based on the grid of bilateral central rates and intervention rates is supplemented by a so-called *indicator of divergence.* It is designed to show whether one of the currencies participating in the intervention system is developing in a markedly different manner from the others, either because of the basic conditions governing exchange rate movements or because it is exposed to special influences. The criterion used is the divergence of the ECU daily rate from the ECU central rate of each participating currency. If a currency reaches a certain threshold of divergence, there is a presumption that measures will be taken to reduce or eliminate the tensions thus indicated in the pattern of exchange rates.

In principle the *threshold of divergence* amounts to 75% of the maximum permissible difference between the ECU daily rate and the ECU central rate of a currency. The maximum permissible difference between the ECU daily rate and the ECU central rate is reached when the market rate of the currency in question against all the other currencies included in the ECU basket diverges by the full margin of fluctuation of 2.25% from the bilateral central rates. (The calculation of the threshold of divergence is based on the assumption that all the currencies included in the ECU basket are participating in the system.) Since the individual currencies have different weights in the ECU basket, the maximum differences between the ECU daily rate and the ECU central rate vary from currency to currency. For a currency with a high weight in the basket the maximum divergence is smaller than for a currency with a low weight. When calculating the indicator the actual divergence of the ECU daily rate of a currency is related to its maximum permissible divergence. The ECU central rates and the divergence limit are given in Table 3.5.

Table 3.5. ECU central rates and divergence limits
(March 1980)

Currency	ECU central rates	Divergence limit %
Belgian Franc	40.7985	±1.5361
Danish Kroner	7.91917	±1.6413
Deutsche Mark	2.54502	±1.1386
French Franc	5.99526	±1.3638
Dutch Guilder	2.81318	±1.5159
Irish Punt	0.685145	±1.6688
Italian Lira	1262.92	±4.1116

When a currency crosses its threshold of divergence, this results in a 'presumption' but not a compulsion that the authorities concerned will correct this situation by adopting one of the following measures:
— Diversified intervention.
— Measures of domestic monetary policy.
— Changes in central rates (as occurred in September 1979).
— Other measures of economic policy.

Supporting Credit Facilities

In carrying out market intervention in support of their currencies, EMS members can use their foreign exchange reserves (primarily dollars) or they can avail themselves of special credit facilities. The special credit facilities have been available to European Community countries participating in the predecessor to the EMS, the snake, but they were expanded to meet the needs of the EMS. These facilities include three types of credits structured by the maturity of the 'loans'.

The first tier consists of almost unlimited amounts of members' currencies that can be borrowed from other participants in the EMS to carry out foreign exchange market intervention. Such loans are available to members for up to 45 days following the end of the month they were made. The loans can be extended, within limits, up to three months. The second tier consists of credits for three to six months, which can be extended to nine months. The amounts that can be borrowed are limited by the size of the pool of credit (about 14 billion ECUs) and by the member's, quota which is determined, in turn, by the relative size of the member's economy. This quota also determines the member's access to the medium-term financial assistance, which is for a term of two to five years. The third-tier pool of funds totals about 11 billion ECUs. However, borrowing under this facility will be conditional on the member's willingness to follow internal economic policies that will reduce the domestic problems that gave rise to the need to borrow.

The European Monetary Cooperation Fund (EMCF)

This institution was set up to administer the various EMS credit arrangements. When a country borrows a currency for intervention, its debt is denominated in ECUs. The debtor country can repay the debt either in the currency it borrowed or in ECUs. A creditor country, however, does not have to accept more than half the repayment in the form of ECUs. The rest of the repayment can be made in the currency borrowed or acceptable international reserves, such as dollars or gold.

Countries that hold more ECUs than their quotas will be paid interest on their excess holdings. Countries that hold fewer ECUs than their quotas will be charged interest on their deficiencies. The interest rate will be equal to the weighted average of the discount rate of the EMS countries. To create an initial supply of ECUs, central banks deposited 20 per cent of their gold and dollar reserves with the EMCF and received an equivalent amount of ECUs. Until establishment of EMCF is formally approved by the legislative bodies of the individual countries participating in the EMS, the deposits will be in the form of revolving three-month swaps.

Chapter Four

THE ROLE AND IMPORTANCE OF GOLD IN THE INTERNATIONAL MONETARY SYSTEM

International Liquidity

By international liquidity is meant the stock of internationally acceptable assets available for the settlement of debts between individuals, corporate bodies and trading nations. These international assets can take various forms, the qualifications for continuous use being that they be liquid, easily transferable, readily acceptable and that they have a stable and predictable value. International liquidity, like national money, serves at least three potential functions: it is a medium of exchange, a unit of account, and a store of value. Individuals and corporate bodies are likely to hold international money to finance trade, for investment and for speculative purposes. Central banks' demand for reserves are likely to be motivated by two conceptually separable forces, a desire to defend the value of their currency on the foreign exchange market and a precaution against the possibility that they will have to finance a balance of payments deficit. Private demand for international liquidity stems from transaction and speculative motives while central banks' demands stem from precautionary motives.

Historically international monies have taken three broad forms; commodity money, fiat money and, credit money. In the context of international trading relations gold has been the principal form of commodity money. As discussed in Chapter 2 commodity monies were used, both domestically and internationally, and among these commodities precious metals, and particularly gold, acquired pre-eminence and eventually drove out competitors. Gold, as a relatively scarce precious metal, was acceptable in international transactions because it fulfilled the function of a medium of exchange and store of value. Over time, however, with a continuous expansion in world trade and a shortfall in gold supplies, the need to economise on existing gold stocks became increasingly apparent. Despite the obvious advantages of paper money (cheaper to produce, more convenient to hold in cash form, etc.) international trust was not sufficient to accept it on its own. Money that has intrinsic value because it is made of a precious metal can be contrasted with fiat money which has value because the public has 'faith' that it will be accepted as legal tender. It

71

was quickly realised that the advantages of fiat money could be reaped if it were backed by something with intrinsic value. Consequently, fiat money backed by gold came to be used in international transactions. The countries whose currencies were used were those which were important trading nations, had a stock of gold with which to back their currency, and were prepared to run balance of payments deficits in order to supply the rest of the world with such liquidity. Thus, prior to the Second World War, the pound sterling was the principal form of fiat money, whereas since the Second World War the US dollar has dominated international monetary relations.

Increasing sophistication of, and confidence in, the domestic economy usually results in an increasing dependence on credit money. This is also the case with the international economy. Thus, over the past decade there have been increased attempts to develop a form of credit money as a further attempt to economise on the use of national currencies. This development has been particularly evident in attempts to increase the role of Special Drawing Rights (see pp. 75-77).

The supply of international money, used between nations, comes from four sources: central bank gold holdings, convertible foreign currencies, IMF drawing rights and Special Drawing Rights.

Central Bank Gold Holdings

As discussed in Chapter 1, the use of gold as a monetary unit is long lived. Following the demise of the international gold standard which had been adopted prior to the First World War the role of gold in international monetary relations went into secular decline. Gold's role at the Bretton Woods Conference (1944), however, was reasserted despite Lord Keynes' protestations about the metal being a 'barbarous relic'. Throughout the 1950s and early '60s, gold's role declined. However, the late '60s and particularly the late '70s have seen a great resurgence in the importance of gold. Following the adoption of the basket measurement of the SDR (see Table 4.2) gold ceased to have any relevance as a unit of account; a state of affairs underscored, in 1976, by the abandonment of the official gold price. Gold remains a reserve asset but its price is allowed to fluctuate with the pressure of the free market supply and demand. Central banks are no longer prevented from buying or selling the metal on the open market. It continues to be used as collateral for international loans and cross guarantees such as that granted to Italy by West Germany in 1974. The reasons why central banks hold gold are discussed in Chapter 5.

Rises in the gold price during the 1970s have induced all the world's central banks, with the notable exception of the USA, to change their gold valuation policies. With effect from January 1975, the gold component of France's international reserves have been revalued at the average price observed in the most representative international markets during the final three months of the preceding half year. Italian official gold holdings are revalued quarterly on the basis of the average London market price prevailing during the 30 day period ending three days before the balancing date, reduced by 15 per cent. With effect from April 1979, the Bank of England's choice of a gold price is to be revised annually, at the end of March, on the basis of the average London market price for gold over the preceding three months, less a 25 per cent discount. In South Africa national revaluation procedures are employed whereby official gold holdings at the end of each month are revalued at the average of the last ten gold fixings recorded during the preceding month on the London market, less 10 per cent.

The gold in countries' reserves at the end of 1979 was 925 million ounces, almost exactly the same as 30 years earlier. The IMF has another 120 million ounces after completing its programme of sales and 95 million ounces was in 1979 switched from the reserves of EEC members into the European Monetary Cooperation Fund. Official gold holdings of Central Banks can be seen from Table 5.7.

Convertible Currencies

When currencies like sterling and the dollar act as international money they are often referred to as key currencies as well as reserve currencies. The motivation for their increasing use is fairly well known. They are less costly to produce than gold, they are more convenient to hold for foreign exchange intervention purposes, they are easier to acquire, and they usually yield income in the form of interest payments.

The mechanism for creating convertible currency reserves differs from that for gold. With gold it is necessary to rely on increased production from South Africa or sales to the West from the USSR. With convertible currency reserves it is necessary to rely primarily on the key currency countries running balance of payments deficits. In other words, the rest of the world runs a balance of payments surplus with the key currency countries by selling them goods and services in exchange for stocks of their currencies which, because of their international acceptability can then be used as reserves.

IMF Drawing Rights

As discussed in Chapter 3, when a country becomes a member of the IMF it is given a quota. This quota provides the basis upon which voting rights and drawing rights are assessed. Each member pays a subscription to the Fund equal to the value of its quota; 25 per cent of this was originally paid in gold, the remaining 75 per cent in the country's own currency. With effect from the ratification of the Second Amendment to the IMF's Articles of Agreement, in 1978, members were no longer required to pay gold to the Fund in connection with any transaction or operation. The

IMF therefore acquires stocks of currencies from its members which can subsequently be lent to nations facing balance of payments difficulties. Any country facing payments difficulties can borrow an amount of foreign currency from the Fund up to the value of its quota. The loan (plus service charges) is repaid over a specified time period (three to five years). These are the country's drawing rights and their importance can be seen by reference to Figure 4.1. The use of the Fund positions forms a relatively small percentage of total reserves.

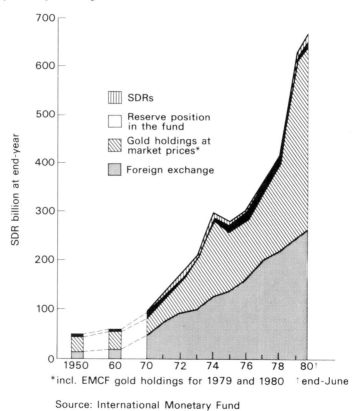

Source: International Monetary Fund

Figure 4.1. Official reserves of IMF member countries.

In addition to these automatic drawing rights any member can further draw on its credit tranche. Here the member can borrow up to 125 per cent of its quota in five slices or 'tranches'. The first drawing (the 'reserve tranche') is unconditional. The remaining four tranches are, however, conditional, the conditions imposed becoming more stringent with each successive drawing. Because of the conditionality element these are not usually included in calculations of countries' foreign exchange reserves. A country's ability to draw on the IMF depends on the IMF's holdings of the drawing country's own currency.

In addition to the reserve and credit 'tranches' other borrowing facilities are available to IMF members. Firstly there is the Compensatory Financing Facility which was devised for countries which for various reasons, such as the weather, suffered temporary set backs in their export earnings. Secondly there is the 'Extended Fund Facility' which was established to assist countries with longer term structural economic problems. Thirdly there is the 'Trust Fund' which uses funds obtained from profits from the IMF's gold sales programme. This is discussed further in chapter five. Fourthly there is the 'Supplementary Financing Facility' also called the 'Witteveen Facility' which is designed to help members to cope with balance of payments problems caused by the rising oil price. Finally there is the 'buffer stock facility' which provides loans for countries to finance commodity stocks.

Following the implementation of the IMF's seventh general increase in quotas, which took effect from December 1980 a country may now borrow in excess of six times its quota (over a three year time period).

Special Drawing Rights (SDRs)

The late 1960s were characterised by increasing uncertainty as to how future increases in the demand for international reserves could be satisfied. It was against this background that the International Monetary Fund decided to create a new international reserve asset. The supply of, and confidence in, this new asset would be independent of any one country's domestic economic policies. The new type of reserve asset which was created to help improve the functioning of the international payments system was the Special Drawing Right (SDR), and came into existence in 1970. SDRs were created as bookkeeping entries and were essentially given to all IMF member countries electing to receive them. These bookkeeping entries were designed to be transferred directly between central banks in settlement of balance of payments deficits, with the IMF originally guaranteeing their value in terms of a fixed amount of gold. Actual holders of SDRs have included only the central banks and Treasuries of IMF member countries which have agreed to accept them, and the IMF itself. Non-IMF members, such as Switzerland, are also allowed to hold SDRs. In recent years an increasing number of bond issues have been denominated in SDRs.

The SDRs are created by the Fund and are issued to members in proportion to their quotas. Any member can draw on its allocation of SDRs if and when it pleases, so long as it maintains an average of 30 per cent of its initial allocation over a five year period. Drawings can then be used to purchase foreign exchange from any other member, since all members of the Fund are obliged to accept SDRs up to a maximum holding of 300 per cent of their initial allocation. Since SDRs never have to be repaid, they contribute a net addition to world reserves. Countries using SDRs are charged a weighted average of interest rates of the most important currencies. Countries receiving SDRs are paid a slightly lower rate. The SDR rates of interest and the rates of remuneration applicable over the six quarters beginning April 1 1979 are as shown in Table 4.1.

Table 4.1. The SDR rates of interest and rates of remuneration: April 1970–June 1980

Calendar quarter beginning	SDR interest rate	Rate of remuneration
April 1 1979	6.50	5.85
July 1 1979	6.75	6.075
October 1 1979	7.75	6.975
January 1 1980	9.25	8.325
April 1 1980	10.25	9.225
July 1 1980	8.25	7.425

Source: 1980 Annual Report of the IMF.

Initially, SDRs were valued at SDR1 = $1. The SDR was originally regarded as 'paper gold' because each unit carried a value equivalent to 1/35 of a troy ounce of fine gold. In order to increase the numeraire role of the SDR (i.e. its role as a unit of account for the international economy) it was vital that the SDRs link with gold was broken. Until July 1974 the SDR was officially valued at 0.888671 grams of fine gold. Thus, in early 1974 the SDR was worth 20 per cent more than the devalued dollar. Since July 1974, in order to achieve this numeraire role, SDR valuation has been on the basis of a weighted 'basket' of currencies. This has thus ended the unit-of-account role of gold. The composition of the basket, which has altered over time, is illustrated in Table 4.2.

Allocations of SDRs can be increased at any time so long as any proposed increase has the approval of 85 per cent of the Board of Governors. SDRs represent, however, only a tiny proportion of international liquidity (*see Figure 4.1*). The allocation of SDRs by the IMF at the beginning of 1978, totalling SDR 4 billion, was the first tranche of an allocation of special drawing rights extending over a three year period and amounting to SDR 12 billion following a decision taken by the IMF in December 1978; the second tranche was allocated at the beginning of January 1980. By January 1981 the total of SDRs in existence reached SDR 21.3 billion.

Table 4.2. SDR basket of currencies.

	July 1, 1974		July 1, 1978		January 1 1981†
	Units of currency included	Weight %	Units of currency included	Weight %	Weight %
US $	0.40	33.0	0.40	33.0	42
D-mark	0.38	12.5	0.32	12.5	19
£ sterling	0.045	9.0	0.05	7.5	13
French franc	0.44	7.5	0.42	7.5	13
Japanese yen	26.0	7.5	21.0	7.5	13
Dutch guilder	0.14	4.5	0.14	5.0	—
Italian lira	47.0	6.0	52.0	5.0	—
Canadian $	0.071	6.0	0.07	5.0	—
Belgian franc	1.60	3.5	1.60	4.0	—
Saudi riyal	—	—	0.13	3.0	—
Swedish krona	0.13	2.5	0.11	2.0	—
Iranian rial	—	—	1.70	2.0	—
Spanish peseta	1.10	1.5	1.50	1.5	—
Austrian schilling	0.22	1.0	0.28	1.5	—
Norwegian krone	0.099	1.5	0.10	1.5	—
Australian $	0.012	1.5	0.017	1.5	—
Danish krone	0.11	1.5	—	—	—
South African rand	0.0082	1.0	—	—	—

†Units of currency unchanged since July 1, 1976.
Source: International Monetary Fund.

How Have International Reserves Grown Over Time?

The 1970s have been characterised by a rapid rise in the amount of international reserves. Reserves had increased from $46 billion at the end of 1949 to $97 billion at the end of 1969, a compound annual increase of 2.75 per cent. In the ensuing years as can be seen from Figure 4.1 reserve growth accelerated. Total reserves, excluding gold, as reported in International Financial Statistics, increased by 11 per cent (SDR 26 billion) in 1979, compared with annual growth rates of 8 per cent in 1978 and about 19 per cent in 1977 and 1976 (*see Table 4.3*). As in past years, the growth of foreign exchange reserves in 1979 accounted for most of the rise in non-gold reserves. However, in contrast to earlier years, official holdings of national currencies rose little (SDR 5 billion). Rather, the 1979 increase in foreign exchange reserves resulted in large part from the issuance, by the European Monetary Cooperation Fund, of European Currency Units (ECUs) equivalent to some SDR 20 billion against deposits of a portion of EMS members' gold holdings. Since March 1979 the members of the EMS have deposited 20 per cent of their official holdings of gold and US dollars with that Fund. The ECUs issued against gold add to the total of

Table 4.3. Official Holdings of Reserve Assets, End of Years 1973–79 and End of May 1980[1].
(in billions of SDRs)

	1973	1974	1975	1976	1977	1978	1979	May 1980
All countries								
Total reserves minus gold								
Fund-relates assets								
Reserve position in the Fund	6.2	8.8	12.6	17.7	18.1	14.8	11.8	12.1
Special drawing rights	8.8	8.9	8.8	8.7	8.1	8.1	12.5	16.2
Subtotal, Fund-related assets	15.0	17.7	21.4	26.4	26.2	22.9	24.2	28.3
Foreign exchange	101.6	126.5	137.3	160.3	200.3	221.1[2]	246.0	258.7
Total reserves excluding gold	116.6	144.2	158.7	186.7	226.5	244.1[2]	270.2	287.0
Gold[3]								
Quantity (*millions of ounces*)	1,020	1,018	1,017	1,013	1,015	1,022	930[4]	934
Value at London market price	94.9	155.1	121.9	117.4	137.8	177.3	361.4	381.5
Industrial countries								
Total reserves minus gold								
Fund-related assets								
Reserve positions in the Fund	4.9	6.2	7.7	11.8	12.2	9.6	7.7	7.8
Special drawing rights	7.1	7.2	7.2	7.2	6.7	6.4	9.3	11.9
Subtotal, Fund-related assets	12.0	13.3	14.9	19.1	18.9	16.0	17.1	19.6
Foreign exchange	65.7	64.9	68.7	73.7	100.0	127.2	135.9	143.3
Total reserves excluding gold	77.7	78.3	83.7	92.7	118.9	143.1	153.0	162.9
Gold[3]								
Quantity (*millions of ounces*)	874	874	872	872	881	884	789[4]	789
Value at London market price	81.3	133.1	104.5	101.2	119.6	153.4	306.7	322.1

Table 4.3. *(continuation).*

	1973	1974	1975	1976	1977	1978	1979	May 1980
Oil exporting countries								
Total reserves minus gold								
Fund-related assets								
Reserve positions in the Fund	0.3	1.9	4.3	5.4	5.4	4.4	3.0	3.1
Special drawing rights	0.3	0.3	0.3	0.3	0.4	0.5	1.0	1.5
Subtotal, Fund-related assets	0.6	2.2	4.6	5.8	5.7	4.9	4.0	4.6
Foreign exchange	10.2	35.0	42.4	49.1	55.2	40.1[2]	51.0	57.6
Total reserves excluding gold	10.8	37.2	47.1	54.9	61.0	45.0[2]	55.0	62.2
Gold[3]								
Quantity *(millions of ounces)*	34	34	35	37	34	36	37	39
Value at London market price	3.1	5.2	4.2	4.3	4.7	6.3	14.2	15.8
Non-oil developing countries								
Total reserves minus gold								
Fund-related assets								
Reserve positions in the Fund	0.9	0.7	0.6	0.5	0.5	0.9	1.0	1.2
Special drawing rights	1.4	1.4	1.2	1.1	1.1	1.2	2.1	2.8
Subtotal, Fund-related assets	2.3	2.1	1.8	1.6	1.6	2.1	3.1	4.0
Foreign exchange	24.9	25.7	25.3	36.2	44.0	52.8	58.1	56.8
Total reserves excluding gold	27.2	27.8	27.1	37.8	45.6	54.9	61.2	60.8
Gold[3]								
Quantity *(millions of ounces)*	110	108	108	101	98	99	102	104
Value at London market price	10.2	16.5	12.9	11.7	13.2	17.2	39.6	42.6

recorded foreign exchange reserves, while the ECUs issued against dollars change only the composition, and not the size, of such reserves.

The volume of gold held in countries' official reserves declined in the 1970s by almost 10 per cent, mainly as a result of the deposits of gold by members of the EMS in the European Monetary Cooperation Fund. The market value of official gold holdings nevertheless doubled between the end of 1978 and the end of 1979 since the decline in volume was more than offset by a rise of 124 per cent in the market price (measured in SDRs). While the market value of official gold holdings rose only moderately in the first five months of 1980, gold valuation adjustments continued to play a major role in raising official holdings of ECUs. The European Monetary Cooperation Fund values the gold deposited by member countries (20 per cent of their holdings) at the lower of (i) the average market price over the six preceding months and (ii) the average market price on the penultimate working day preceding the swap period for which the ECU price of gold holdings is to be established. According to this formula, the price of gold used by the EMS rose from ECU 165 per ounce for the first swap period started March 13, 1979 to ECU 211 for the third swap period ended January 8, 1980. During 1979, the resulting revaluation of the stock of gold deposited, whose volume remained essentially unchanged at 85 million ounces after July 1979, amounted to almost SDR 5 billion. Further gold revaluations, to ECU 371 per ounce, added almost SDR 15 billion to foreign exchange reserves during the first five months of 1980.

Valuing official gold holdings generally at the free market price poses problems for the international financial community, some of which are discussed below. On the basis of such valuation enormous inflation of

Notes to Table 4.3.

Source: *International Financial Statistics.*

[1] 'Fund-related assets' comprise reserve positions in the Fund and SDR holdings of all Fund members. Claims by Switzerland on the Fund are included in the line showing reserve positions in the Fund. The entries under 'Foreign exchange' and 'Gold' comprise official holdings of the Netherlands Antilles, Switzerland, and Fund members except the People's Republic of China, for which data are not published. Figures for 1973 include official French claims on the European Monetary Cooperation Fund.

[2] Beginning with April 1978, Saudi Arabian holdings of foreign exchange exclude the cover against the note issue, which amounted to SDR 4.3 billion at the end of March 1978.

[3] One troy ounce equals 31.103 grams. The market price is the afternoon price fixed in London on the last business day of each period.

[4] The decrease recorded in the quantity of countries' official gold holdings from the end of 1978 to the end of 1979 reflects mainly the deposit by the nine member countries of the European Monetary System of 20 per cent of their gold holdings with the European Monetary Cooperation Fund. The European Currency Units (ECUs) issued in return for these deposits are shown as part of the countries' official foreign exchange holdings.

Table 4.4. Components of International Liquidity.

	1960	%	1978	%	1979	%
SDRs	0	0	10.5	2	10.6	0.5
Reserves at IMF	3.6	5	19.3	3	19.4	1.5
Foreign Exchange	18.4	29	287.1	50	289.3	26
Gold	40.9	66	258.6	45	802.7	72
Total Reserves	62.9	100	575.3	100	1122.0	100

(Derived from *International Financial Statistics*.)
Note: Gold is valued at $700 per ounce for 1979.

international liquidity would result for 1979. As can be seen from Table 4.4, if gold is valued at $700 per ounce for 1979 this would imply a growth in the dollar value of worldwide official reserves of some $540 billion. This is a considerably greater addition to world reserves than that which occurred from all non-gold sources over the whole period 1960–1978. While international reserves grew by an average of 10 per cent per year from 1960 to 1978 the increase between 1978 and January 1980 was nearly 100 per cent. The share of gold at the $700 valuation rate contributed 72 per cent of total reserves in 1979 compared with a figure of 45 per cent in 1978.

There are several reasons for not valuing monetary gold at its free market price when assessing the adequacy of global international liquidity. First and foremost among them is the fact that the gold price is subject to wide fluctuations induced by speculation and that the gold markets are relatively narrow; these reasons taken together imply a high liquidity risk. Against this view one could argue that gold is only unstable in relation to paper currencies which are themselves highly volatile, both in relation to each other and in terms of their purchasing power. Paper currencies have performed erratically over the past decade and in terms of conventional yardsticks some have gone up, some down, but all have crashed against gold. This aspect of the gold market is discussed further in Chapter 6.

What Characteristics Should a Reserve Asset Possess?

Having discussed the major sources of reserve assets it is important to assess what characteristics a reserve asset should possess. Williamson has stressed that the suitability of a reserve asset should be assessed with regard to three broad considerations: 'confidence', 'stabilisation', and 'seigniorage'.[1] The confidence problem arises from the simultaneous use of more than one form of reserve asset. In these circumstances whenever doubts arise regarding the acceptability of any one asset, the possibility exists that reserve holders (both official and non-official) may substitute that asset for another placing strains on the system. Given the post war strength of the US economy and its huge gold stocks, the US dollar was regarded as being 'as good as gold'. US dollar liabilities were after all backed by gold on a greater than one to one basis (*see Figure 3.2*). Countries therefore accumulated dollar reserves supplied through a US balance of payments deficit. Since, however, the US was under no obligation to maintain a fixed relationship between its gold assets and its dollar liabilities, her liquidity position steadily worsened. Eventually, this caused creditors to become suspicious of the ability of the United States to meet its obligations.

This suspicion was manifested in two ways. First, several central banks converted their dollar holdings back into gold (which naturally exacerbated the US liquidity position by worsening the asset:liability ratio). Second, private speculators, seeing the possibility of speculative capital gains based on a dollar devaluation, sold dollars in exchange for other currencies. A dollar devaluation would have been a natural response for the authorities to undertake given the need to eliminate the US balance of payments deficit. An important lesson which emerged from these developments in the 1960s was that where an international monetary system was based on more than one type of reserve asset, problems of confidence in the system could eventually emerge. It is for this reason that some commentators, notably Jacques Rueff, have argued that there should be a single form of international money (namely gold) which would then eliminate the possibility of reserve asset substitution.

This solution would no doubt solve the confidence problem but as stated earlier, Williamson had stressed that two other considerations are important: stability and seigniorage. By stability is meant the ability of the supply of international money to facilitate a sustained growth in world trade. This supply ought not to be excessive otherwise world inflation may be generated and fuelled. Nor should it be deficient, in which case deflationary tendencies may predominate. Ideally the supply of international money should change in a counter-cyclical manner in order to stabilise the world economy. From the stabilisation point of view the liquidity arrangements of the post war Bretton Woods system seem to have been inadequate. With the 'official' price of gold being held at $35 per ounce, and mining costs rising, gold supplies for monetary use increased very slowly. The supply of international liquidity thus became primarily dependent on convertible currencies, principally the dollar, and thereby on the size of the US balance of payments deficit.

Another consideration in developing an international reserve asset is the seigniorage problem. As discussed in Chapter 2 commodity monies have been supplanted by paper monies in all advanced national states, the reason being that paper money can be created at very little cost. The saving achieved by using paper money instead of commodity money can be approximated as the value of resources that are freed from the need to create the increase in the stock of the commodity used as money. If a central bank has a monopoly over the printing of money it can obtain a monopoly profit which is approximately equal to the difference between the resources used in creating the commodity money and the resources used in printing the paper money. This monopoly profit is known as seigniorage. If the world were on a gold standard, increases in the world money supply could only be achieved by the surrender of real resources to the gold mining industry. Thus under a gold standard the problem of seigniorage does not occur. However with a reserve currency system the country issuing the reserve currency is able to extract seigniorage from the parties holding the reserve currency. Any country wishing to increase its stock of international reserves may have to do so by running a payments surplus with the reserve issuing country (i.e. the USA or in the past the United Kingdom). In other words, a country has to export goods, i.e. give up the domestic use of real resources, in order to obtain paper money, the rate of return on which may be either zero, very low, or even negative. Many countries have objected to this seigniorage accruing principally to the world's richest nation, viz. the USA. President de Gaulle called this property of the system 'an exorbitant privilege' for the Americans. As already mentioned the seigniorage problem would be resolved by the use of a commodity money such as gold. However presumably it is also desirable to reap the available resource saving, mentioned earlier, involved in substituting paper money for commodity money.

America versus France:
'The marvellous secret of the deficit without tears': contrasting views on the role of gold

Some flavour of the arguments in favour and against gold remonetisation, developed further in the next two sections, can be gathered by analysing the contrasting views of the French (long in favour of gold remonetisation) and the USA (long against gold remonetisation).

By the mid 1960s there was widespread acceptance of the proposition that for industrial countries exchange rate changes, particularly for the dollar, should be infrequent occurrences. (A rise in the price of gold is equivalent to a devaluation of the dollar.) There were various reasons for this view. First, there was an almost universal belief that the fixed price of gold at $35 an ounce was a basic underpinning of the system. To almost everyone, this ruled out unilateral devaluation by the United States. According to Theodore Sorensen, some of President Kennedy's advisers privately told him that even devaluation was not unthinkable, that it would be a drastic change in the system, but preferable to wrecking it altogether.[2] The President responded that he did not want that weapon of last resort even mentioned outside his office; by disrupting the international monetary system, devaluation would call into doubt the good faith and stability of the nation and the competence of its President.

Aside from questions of 'good faith' and prestige, there was a seemingly overriding economic reason to oppose a devaluation of the dollar involving a rise in the price of gold. This was that so much of the world's foreign exchange holdings consisted of dollars which were retained by monetary authorities on the assumption that the dollar would maintain its value in relation to gold. If the dollar price of gold were changed, the entire monetary system would have to be altered, for it could not be expected that foreign monetary authorities would go on holding dollars in the belief that the dollar's gold value would not be changed again.

Other reasons why the US opposed the revaluation of gold (i.e. raising the official gold price) were given by Triffin.[3] First of all, the increase in gold prices needed to stimulate adequate annual supplies of monetary gold would have to be very steep indeed. Such supplies would probably have to be doubled or tripled – in dollar terms – in order to finance the needed growth in international liquidity. Second, this gold revaluation operation would have to be repeated at periodic intervals, in order to keep pace with the cumulative growth of the world economy. Third, each of these revaluations would result in a temporary excess of world liquidity, due to their impact upon the valuation of existing gold reserves at the time. Fourth, the benefits of gold revaluation would be distributed very haphazardly and, indeed, in just about the least desirable fashion imaginable. High reserve countries would benefit most, and low reserve countries would benefit least, from the revaluation of existing stocks. The USSR would undoubtedly be the major beneficiary in this respect after the United States. It would, moreover, be the largest beneficiary by far, together with South Africa, of the consequent rise in the dollar value of current gold production. Last but not least, the world liquidity requirements could certainly be put to better use than the financing of more and more earth digging in South Africa, the USSR, Canada, the United States and Australia.

The dilemma of the gold exchange standard as posed by Triffin in 1960 was that either liquidity would become progressively more inadequate as elimination of the United States deficit restricted liquidity growth to the net supply of new gold, or else a continued deficit would lead to a progressive deterioration in the United States reserve ratio, thereby undermining confidence in the dollar and eventually provoking massive attempts to convert dollars into gold, with the result that the system would again collapse as it did in 1931.[4] Thus continuing US deficits were both necessary yet alarming.

France's role as maverick of the international monetary system has become familiar to two generations. In early 1965, the French Government decided to convert into gold some $300 million of its dollar holdings and thereafter to step up its monthly purchases of gold from the United States. These actions prompted a question at one of President de Gaulle's magisterial press conferences, on February 4, 1965. He was asked whether he favoured a reform of the international monetary system and, if so, how.

The lengthy reply brought out the by-then well-known French criticisms of the existing system, notably the privileged position of the USA, which permitted amongst other things 'a kind of expropriation' of enterprises in other countries. The French were particularly concerned with the freedom that the IMF system gave the United States to invest in foreign countries. Assume that an American company wanted to acquire a French factory. It would sell dollars and buy francs to pay for the factory. In order to keep the exchange rate pegged, the French authorities would have to enter on the other side of the market buying the dollars and selling the francs being demanded by the United States company. In other words, the IMF system required the French authorities to provide the francs with which United States companies could finance their takeover of French industry. To the French, this was ridiculous. In this period of relative calm, de Gaulle pointed out it was opportune to change the system and establish international trade:

'On an unquestionable monetary basis that does not bear the stamp of any one country in particular. On what basis? Truly it is hard to imagine that it could be any standard other than gold, yes, gold, whose nature does not alter, which may be formed equally well into ingots, bars or coins, which has no nationality and which has, eternally and universally, been regarded as the unalterable currency par excellence. . . Certainly the terminating of the gold exchange standard without causing a hard jolt and the restoration of the gold standard as well as the supplementary and transitional measures which will be essential, particularly the organizing of international credit on this new basis – all that must be examined calmly.'

This pronouncement was something of a bombshell in financial circles. In substance it did not diverge from the position French officials had been promoting, but it was generally interpreted as a call for a further step towards a full gold standard. Given the authority from which these words came, the reactions were to be expected. The US treasury issued a statement on the same day rejecting a return to the gold standard and re-affirming the US intention to maintain the official price of gold. In London, *The Economist*, which has long opposed gold remonetisation, expressed fears that by President de Gaulle's evidence, one could infer that the influence of Jacques Rueff had gained the upper hand over that of officials in the Ministry of Finance and the Bank of France.

While sharing substantially Triffin's analysis, Jacques Rueff, General de Gaulle's architect of French economic and monetary stabilisation, did not share the conclusion.[5] Rueff saw the world repeating the gold economising policies adopted in the 1920s whereby countries gaining sterling or dollars did not convert them into gold as they would have done under a genuine gold standard. They left the funds on deposit in the countries of origin; the international monetary system became a 'childish game' in which the winners returned their marbles to the losers: the key currency countries learned 'the marvellous secret of the deficit without tears'. Balance of payments adjustment ceased being quasi-automatic and came to depend on deliberate measures of credit or trade control policy. Growing claims of surplus countries on deficit countries could serve as a basis for credit expansion in the former without imposing contraction on the latter (which the 'rules' of the gold standard demanded). But this duplication of the credit pyramid on a relatively narrow gold base could not go on indefinitely. To use a further metaphor, Rueff stated that 'if I had an agreement with my tailor that whatever money I pay him he returns to me the very same day as a loan, I would have no objection at all to ordering more suits from him'. Rueff went on to claim that the house of cards was getting shakier and would eventually collapse as in 1929–31. The danger had developed not because the USA had lost so much gold since 1950, but because it had lost so little, delaying adjustment of the balance of payments.

What Rueff was aiming for was the restoration of a system whereby the debtor country loses what the creditor country gains. Rueff was striving to achieve the restoration of the rules of the gold standard. Thus central banks should create money only against gold and should not build up assets in dollars which are then re-lent to the USA via the purchase of US Treasury Bills, thereby automatically refinancing the US balance of payments. As part of the means to achieve this, Rueff insisted that it would be necessary to raise the price of gold. The reason for this was that the overhang of dollar balances was so great that they must be either funded or reimbursed. Funding is the conversion of short-term debt into longer term, more permanent obligations. Sales of long dated securities, coupled with purchases of short dated securities, tend to reduce the liquidity base of the banking system and consequently would be very deflationary. Reimbursement could be either through the Triffin plan or by raising the gold price. Under the Triffin plan (superseded by the proposed substitution account) the dollar balances would be placed into IMF deposits, a scheme distrusted by Rueff. Rueff favoured the rise in the gold price which would eliminate the existence of the dollar balances enabling them to be paid off once and for all.

As Bergsten has pointed out, the crucial requirement of any return to the gold standard, which may be needed in order to achieve a better adjustment process is international agreement that national monetary authorities would never again add dollar (or other national currency) balances to their reserves.[6] Only such an agreement would result in the certain elimination of the ability of key currency countries to finance payments deficits by increasing their liabilities, the means by which proponents of this approach would hope to achieve a better adjustment mechanism for the system as a whole. The reserve centres could contribute to the implementation of any undertaking to end their reserve currency roles by erecting controls on capital flows. In a debate with Triffin and Bergsten in 1967 Rueff stated that he did not rely on automatic adjustment to result from adoption of the gold standard, but rather on the incentives to countries to adjust if they knew that they would have to finance any deficits by selling gold.

Bergsten goes on to show that from the standpoint of this adjustment objective, it would in fact be irrelevant whether the dollar overhang was eliminated or not, since the crucial consideration would be the agreement to avoid future increases in dollar reserves. Thus the proposed increase in gold price would not be an integral element of a 'return to the gold standard'. Furthermore, the world could agree to avoid any future build-ups of reserve currency holdings at the same time as it was demonetising gold, either through turning completely to SDRs or to freely flexible exchange rates. And it could eliminate the dollar overhang by completely substituting a truly international asset for it. In fact any sizeable increase in the price of gold without agreement on future dollar build-ups could revive the reserve role of the dollar as indeed happened after 1934. Such a step would drastically improve all of the US liquidity ratios, thereby improving international confidence in the dollar. Thus an increase in the gold price without a corollary agreement on future dollar balances would fit into the option for increasing (and not decreasing) the reserve role of the dollar!

Thus Bergsten claims the entire Rueff proposal is inconsistent on two fundamental points: its adjustment objective could only be achieved by international agreement to eliminate the reserve currency roles of the dollar for the future, and could even be undermined by any proposed increase in the gold price in the absence of such an agreement; and there would be no need to convert the dollar overhang into gold, and hence to raise the official price to permit such a, payoff, once the basic agreement was reached to avoid future dollar build-ups. As Bergsten puts it: 'In short, "gold standard" adjustment could be achieved without reliance on gold for liquidity, and gold could be relied on for liquidity without the adoption of "gold standard" adjustment rules.'

Should Gold's Official Role in the International Monetary System be Reduced (Demonetisation)?

Attempts to demonetise gold are not new. According to Plutarch the first historical attack on the role of gold was by Lycurgus (c. 825 BC) in Sparta. Lycurgus introduced a monetary system under which the State authority made it compulsory to employ as media of exchange iron bars rendered useless for practical purposes by a special process. The possession of coins or precious metals was outlawed and was subject to severe penalties. Only the Government itself was permitted to own coins which were used exclusively for the requirements of foreign trade or other payments abroad. Another well-known historical experiment, much nearer our time, was that of John Law in France during the early part of the eighteenth century. It coincided more or less with the South Sea Bubble in England and with the frenzied speculation in tulip bulbs in the Netherlands. John Law persuaded the Regent's Government to authorise the issue of paper money. In order to increase confidence in the currency of his invention, Law caused the official price of gold coins to be changed frequently in the hope of conveying the impression that the value of the coins was unstable and that therefore they were undependable. The experiments of both Lycurgus and Law failed.

In the nineteenth century John Ruskin argued furiously against gold. During the economic depressions of the inter-war period, Keynes and McKenna, the United Kingdom Chancellor of the Exchequer, were at the spearhead of the movement aiming at terminating the monetary role of gold. They were opposed to the gold standard on the ground that the discipline it had imposed on the economy had prevented a credit expansion which would have reduced unemployment. It was not discovered until the publication of the Keynes papers in Volume XIV of his collected works in 1971, that during the First World War both he, in his capacity of economic advisor to the Treasury, and McKenna as Chancellor of the Exchequer, were firmly convinced that unless the pretence of Great Britain being still on the gold standard was upheld, her prestige and her financial power would decline to such an extent as to make it impossible for her to continue fighting Germany. The views of Keynes and McKenna represented the spirit of the times. Their change of attitude after the war was the result of inter-war depression calling for an expansionary monetary policy, the adoption of which was prevented by the need for credit restraint in order to maintain the gold standard on the basis of the 1914 parity of sterling. The extent of the change of attitude adopted by Keynes can be seen from a section on his article entitled *Auri Sacra Fames*, published in *A Treatise on Money* (*page 179*) in September 1930:

'In truth, the gold standard is already a barbarous relic. All of us, from the Governor of the Bank of England downwards, are now primarily interested in preserving the stability of business, prices and employment, and are not likely, when the choice is forced on us, deliberately to sacrifice these to the outworn dogma, which had its value once, of £3 17s. 10½d. per ounce. Advocates of the ancient standard do not observe how remote it now is from the spirit and the requirements of the age.'

The fact that the rate of production of gold has usually been low compared with the total stocks of gold in circulation (*discussed further in Chapter 5*) has frequently been cited as a favourable attribute of gold in its monetary role. It must be pointed out that this factor was favourable only when gold was used as currency and may actually be unfavourable when gold is utilised as a monetary reserve rather than as currency itself. As long as gold was used as the primary form of currency, physical limitations on the potential for expanded new production were important to prevent inflation. The 'quantity theory' in the formulation by the American economist Irving Fisher is a useful tool for dispelling the apparent paradox as to whether this attribute is favourable or unfavourable. The Fisher equation, quantity of money multiplied by velocity equals the price level multiplied by the physical volume of transaction, must hold as an identity in the static sense, but the question has been actively debated as to what functional relationships can be derived in the dynamic sense. It is a truism in the sense that, if the quantity of money in circulation were to change drastically and quickly without simultaneous offsetting changes in the velocity of circulation and the physical volume of transactions, the price level must rise or fall, in line with the change in the quantity of money. In this context, relative inflexibility in the quantity of gold makes sense as a positive attribute contributing to stable prices when gold is used as currency or as a determinant of the supply of money. However, in the current situation, where the stocks of gold are held privately and as monetary reserves rather than as currency in circulation, the above argument is no longer applicable. Changes in the quantity of gold held as monetary reserves need not directly influence the price levels, and a limit on the physical quantity of gold available for use as monetary reserves does not serve as a direct restraint upon inflation. The fact that gold is no longer used as currency or as a determinant of the supply of money means that the positive aspect of this attribute has been considerably weakened.

Williamson has listed several considerations which need to be taken into account when considering the merits of gold compared with the Special Drawing Rights as the basic international reserve asset.[7] First in an inflationary age there is no question of a once-for-all gold revaluation leading to a permanent solution of the need for steadily growing liquidity such as can be provided through regular SDR allocations at a controlled rate. The crux of the case against gold lies in the element of inflexibility it introduces into the control of the world's supply of liquidity. Since the existing flow of gold production is very small in proportion to current monetary stocks, and production is relatively inelastic with respect to price, the physical stock of monetary gold can be regarded as more or less given in the short or medium term. South Africa's annual production, which is about 73 per cent of non-communist world output, represents a little over 2 per cent of the world's monetary gold stock. The scope for increasing monetary gold from new production is therefore rather limited. There is somewhat more elasticity in the supply of gold in private hands not currently devoted to monetary use. In principle, some of this hoard could be tempted into official hands, but in practice it has proved difficult or impossible to achieve this. But even if it were possible to attract gold from private hands into monetary uses, it would not really be satisfactory to have liquidity supply arrangements in which liquidity moved in and out of the monetary system in response to largely capricious changes in speculators' demands.

Second, in reply to the claim that gold reinstatement would restore financial discipline on governments Williamson points out that once governments learn they can change the gold price whenever it becomes an irksome constraint they will never be fully disciplined by a gold standard system. Third gold mining absorbs real resources in terms of mining costs whereas SDRs can be created out of thin air. Thus gold could be released from monetary uses and used for filling teeth and for ornamental uses. With positive gains accruing from the substitution of SDRs for gold then a competitive interest rate could be paid on SDRs which may enable them to be used as currency alternatives by reserve asset holders thereby avoiding disruptive switches between currencies.

Fourth with restrictions being imposed on the switching possibilities when holding SDRs the confidence problem which could occur with any resurrected gold standard would be reduced. Fifth the sudden massive increase in international liquidity that would occur whenever gold was revalued would have highly inflationary effects. This problem occurs however even if gold is not remonetised but simply rises in price in line with market forces. (*See pp. 95-100 of this chapter for an outline of the monetary consequences of the rise in the gold price.*)

Finally, the distribution of the seigniorage benefits of SDR creation is at least roughly neutral, and could, with sufficiently general agreement, be varied to reflect any consensus that might be established on a desirable international redistribution of income. The distribution of the seigniorage benefits of gold revaluation would have been at best capricious and at worst perverse, with the major gold producers (South Africa and the Soviet Union) and the gold hoarders benefitting substantially, and most of the developing countries gaining virtually nothing.

In conclusion then there are powerful arguments against the use of gold as compared with the use of SDRs.

Should Gold's Official Role in the International Monetary System be Increased (Remonetisation)?

These defects of gold as a reserve asset as compared with SDRs are well recognised, although the advocates of gold as an international reserve asset would minimise their importance. They would say that although gold-mining is an unproductive activity in so far as monetary gold does not satisfy intrinsic wants, the proportion of the world's resources currently devoted to gold-mining is trivial, less than one tenth of one per cent. This

could be considered an acceptable price to pay if the international economic system were more efficient as a result. Similarly, the arbitrariness of
the real income transfer resulting from gold-mining is also recognised; but
this objection applies to the exploitation of all raw materials, and in a
world where the price of oil (a vastly more important component of world
trade) can be quadrupled in a few months, nobody would pay much attention to the income transfer aspects of gold were it not for the political fact
that the two principal producers are South Africa and the Soviet Union.

A common argument for gold remonetisation is based on the need for
increased international liquidity. This necessity occurs for two reasons.
First, with worldwide inflation the simple rise in prices increases the
need for more monies in order to undertake the same amount of international transactions. Second, as the world's capital and goods markets
have become more integrated, which in turn increases the volume of
international transactions, there is a necessity for an increase in the amount
of international money. These two factors act as multipliers on the volume
of financial transactions and thus on the demand for international liquidity. Only the dollar or gold can act as the basis of an international currency system as there are simply not enough of any of the alternatives.
Despite the increased willingness of Germany, Switzerland and Japan to
increase the importance of their own currencies these still represent only
a small, albeit increasing, percentage of foreign exchange reserves. In rough
terms the dollar accounted for 85 per cent of all official currency holdings
in 1971; by mid-1980 that share had fallen to about 75 per cent. The USA
has historically shown increased reluctance to accept the disciplines on
domestic policy necessary to maintain the status of the dollar although the
election of Ronald Reagan may alter these trends. This reasoning leads to
the conclusion that it is to gold remonetisation that one must look to provide the means for increasing international liquidity.

The arguments against gold remonetisation have been paid lip service
to by central bankers while in practice these same central bankers have
been very reluctant to give up their gold reserves. Their reasoning is that
gold is beyond the control of any one nation. As a result gold makes it
easier than any alternative arrangements, whether an inconvertible dollar
standard controlled by the USA or money controlled by an international
body, for governments and central banks to safeguard a substantial measure of autonomy and independence in their own monetary and fiscal
matters.

Another unique quality of gold is that there are times and circumstances where no other asset is a substitute because gold alone is universally acceptable, without any questions being asked, as a payment of last
resort. This motive is equally true for a central bank fearing the blocking
of its dollars in the USA as it is for the Vietnamese boat people fleeing
from political persecution. The decision taken by President Carter, in the
wake of the seizure of the US embassy in Tehran in November 1979, to

freeze about $8 billion of Iran's assets in US banks (a decision reversed in January 1981), including more than 50 tonnes of gold held at the New York Federal Reserve Bank, has had important consequences for the gold markets. This blocking action has exerted a powerful influence on Iran and Iraq, who went to war August 1980, as well as on other countries. It has encouraged a number of countries, Iran is only the main example, to diversify further their reserve holdings into non-dollar currencies and gold. It has also made the Gulf war countries among others more eager to hold their assets in places deemed more secure from interference: currencies have been moved to non-US banks and gold has been shifted back home from foreign depositories.

Customs statistics from the United Kingdom and Switzerland, the two centres of the international bullion market, throw some light on what has been going on. They show that between autumn 1979 and October 1980 Iraq has transferred about 90 tonnes of gold, worth $1.8 billion from Zurich to Baghdad. Iran has brought home a total of 46 tonnes from both centres, worth nearly $1 billion. A number of other oil exporting countries have also shifted large amounts of bullion from Zurich and London. In common with normal international practice Iran's gold reserves were spread around the world at the time of the freeze with about 40 per cent of its published holdings on deposit with the Federal Reserve Bank of New York. This gold, worth roughly $1 billion, was blocked between November 1979 and January 1981 and this helps explain the alacrity with which Iran retrieved gold held elsewhere in the West since the asset freeze. The bulk of the 29 tonnes of Iranian gold brought back to Tehran in 1980 from London is thought to have been lodged with the Bank of England.

Advocates of a return to the gold standard often point to historic experience in order to illustrate that there has been a close relationship between the gold standard and price stability. As Rees-Mogg has illustrated, in the United Kingdom, from the restoration of King Charles II in 1660 down to the outbreak of the First World War, Britain operated on an unqualified gold standard.[8] During the pre-Napoleonic period from 1661 to 1796, according to a table of prices compiled by *The Economist*, the low point of prices occurred in 1743, at an index figure of 62 against 100 for 1661 (*see Table 4.5*). The high was 114 in 1796, shortly before the suspension of convertibility. In the post-Napoleonic period, from 1822 to 1913, leaving out 1797 and 1821 when there was not convertibility for the whole year, the low was again 62 in 1895, and the high was 115 in 1825, again close to the non-convertible period. The high of the wartime inflation of the Napoleonic Wars was 180. There must be some reservation about price indices compiled over such a long period, but the price indices do indicate a high degree of stability for the two hundred and fifty years after the Restoration, with the exception of the Napoleonic period.

This long period of history cannot be dismissed as uneventful. Although it excludes the Napoleonic War, it includes Marlborough's wars; it includes

Table 4.5. Prices since 1661 (1661=100)*.

Figures for each decade are along a horizontal line: e.g. that for 1667=88, for 1944=195.

	0	1	2	3	4	5	6	7	8	9
166—		100	103	101	96	96	92	88	88	84
167—	85	84	81	80	86	92	88	81	82	87
168—	85	82	82	80	81	83	84	74	74	73
169—	75	76	75	78	87	87	89	90	95	98
170—	85	74	73	70	73	66	75	65	68	79
171—	90	100	75	72	76	77	73	70	69	72
172—	75	74	68	66	70	72	75	71	73	77
173—	70	65	66	63	65	66	64	69	67	66
174—	74	80	73	70	62	63	69	67	70	71
175—	70	67	69	67	67	68	68	81	78	74
176—	73	70	70	74	75	78	79	81	80	73
177—	74	79	87	88	86	84	84	80	87	82
178—	81	85	86	95	93	89	88	87	90	87
179—	92	90	90	95	101	109	114	110	110	118
180—	157	169	129	115	119	138	136	138	151	157
181—	153	152	175	180	155	141	127	140	144	128
182—	115	105	101	104	106	115	102	102	97	95
183—	95	97	95	93	97	97	107	102	103	113
184—	111	105	96	91	94	95	95	100	87	82
185—	82	79	82	97	108	108	108	110	96	100
186—	104	100	104	105	103	102	104	102	100	93
187—	95	100	111	110	105	102	100	95	88	85
188—	89	86	88	88	82	76	72	70	73	73
189—	76	75	71	71	64	62	63	64	68	73
190—	79	75	75	75	72	75	81	84	76	80
191—	84	86	91	91	91	116	146	193	207	222
192—	270	167	141	139	150	146	136	131	129	124
193—	104	89	86	85	103	103	106	110	113	113
194—	152	205	195	177	195	191	191	205	219	227
195—	234	251	269	273	276	287	301	312	319	319
196—	322	333	347	354	365	379	396	404	425	446
197—	474	513	545	595						

*Linked index. Main sources: Mitchell and Deane, Abstracts of British Historical Statistics, and Department of Employment, British Labour Statistics Historical Abstract 1886-1963.
Basic series: Schrumpeter-Gilboy price index 1661-1697 (1697=100) and 1696-1823 (1701=100); Rousseaux price indexes 1800-1923 (1865 to 1885=100); Sauerbeck Statist price indexes 1846-1938 (1867 to 1877= 100); DE index of the internal purchasing power of the pound 1914-1968 (1969=100).
Series rebased on 1661=100 using the multipliers: 1697=100; 0.9174; 1701=100, 0.7399; 1865-1885=100, 0.8679; 1867-1877=100, 1.0761; 1963=100, 3.5417.

Source: *The Economist*, July 13, 1974.

one successful and three unsuccessful attempts to change the British regime; it includes the industrial revolution; it includes the impact of the American War of Independence and the American Civil War; it includes the creation of the British Empire, virtually the whole of it; it even includes the creation of the trade unions, and it includes the development of modern parliamentary democracy based on mass suffrage. Nor was it a period of long-term economic depression or of failure to increase wages. Industrially it was a most creative period in British history and real wages rose firmly, though not without interruption.

The problem with statistical evidence which correlates two variables is always the one of causality; do the gold discoveries precede rising prices and gold shortages precede falling prices, or does causality run in the other direction? Channels connecting gold and price movements are many but some of these do not operate very clearly. Thus the observed historical connection is not easy to translate into a causal mechanism. In the nineteenth century, as the examples of Klondike and California show, gold discoveries depended on pure chance. This was not always the case. As Vilar has shown, the search for gold, in Peru in the sixteenth century or Brazil in the seventeenth, was not based on precise economic calculations of comparative profitability.[9] Nonetheless the more precious metals prices rose relative to other commodities as a whole, the more fervently they were sought. The great discoveries always took place in periods of generally low prices, in other words of very high relative prices for gold and silver. Christopher Columbus did not emerge by pure chance. The discovery and sudden exploitation of undreamt-of wealth, at first often to be had for next to nothing, determined sudden changes in the relative prices of gold or silver and commodities, initially on the spot (Peru in 1534 and California in 1849), and then in the longer term, in the countries most closely connected with the gold-producing areas. One therefore has to look, not for some unilateral causation but for historical causation which combines chance with necessary mechanisms of change.

In the USA, in the period 1879 to 1914, the dollar was convertible into gold at a fixed ratio specified by law and maintained in practice. Figure 4.2 shows the high degree of stability of wholesale prices in that period.

A problem with correlating changes in gold production and world prices is that gold was never the only form of money. Even in the sixteenth century more transactions were paid for in the account books at the great fairs than were paid in gold and silver; only the balances were paid in precious metals. The history of money has long been characterised by the substitution of credit money for commodity money.

The fact that since leaving the gold standard many countries have suffered monetary inflation must be cautiously interpreted. The monetary inflation that has taken place may not be a result of the ending of the link with gold. The inflation may simply highlight underlying forces that were caused by other factors, e.g. the price of energy, the increasing role

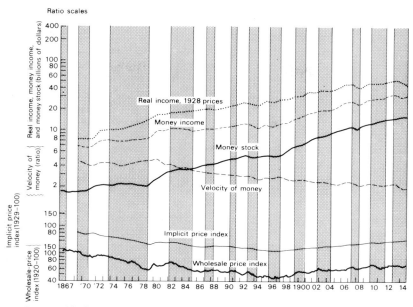

Source: M. Friedman and A. Schwartz *A Monetary History of the United States*

Figure 4.2. Money stock, income, prices and velocity, in reference
wide expressions and contractions, 1867–1914.

of governments, etc. that have nothing to do with the gold link. Causality
is always a problem in economics.

A final last word on the arguments for and against gold remonetisation
comes from Janos Fekete, Vice-President of the Hungarian National
Bank. Fekete commenting in 1972, but still as applicable today:

'There are three hundred leading economists in the world who are
against gold. The problem is that these three hundred scholars are
opposed by three billion people on earth who still believe in gold as
much as ever.'

The Monetary Consequences of a Rise in the Gold Price

What then are the potential monetary consequences of the recent rise
in the gold price? In normal circumstances, increases in a country's reserves are financed by the sale of short-term government debt to the banking system, generating a potential increase in the money supply via the
bank credit multiplier. Thus in order to supply its own currency in exchange for foreign currency the central bank would normally have to

finance this activity by the issue of short dated debt such as Treasury Bills. Where the value of the existing portfolio of reserves is raised by means of a simple balance sheet revaluation, the resulting 'profit' reduces the need for treasury bill sales, thereby neutralising the effect on the money supply. For those economies with large holdings of gold the windfall addition to reserves could, if not already discounted, make it easier for them to market longer term paper thus either reducing interest rates or reducing the money supply. It follows then that there are no automatic effects of the gold price rise, as has sometimes been claimed, which would automatically lead to an increase in a country's money supply.

Some governments may choose to use the reserves windfall to expand their economies or to delay planned fiscal cuts. This might be relevant in countries such as Belgium, Italy, Portugal and Turkey and more particularly in the developing countries which have significant gold reserves (e.g. Argentina, Colombia, Egypt and Algeria). Most countries, however, regard their domestic policies as being constrained by the need to contain or squeeze out inflation so the above tendency will probably not be widespread, at least within OECD countries.

The principal beneficiaries of the increase in gold values have been the industrialised economies who between them hold 83 per cent of the total in official reserves. (Table 4.6 shows that USA, France, Germany, Italy and Switzerland account for 64 per cent of this total.) Most of these countries have substantial holdings of other exchange reserves and are unlikely to change their monetary/fiscal stance simply to exploit the extra wealth the gold price increase gives them. However, the windfall gain to central banks is very large in some cases, and there will undoubtedly be some incentive to run down the level of the reserves in some countries.

While some central banks may seek to sell gold to other central banks in return for other resources, for the world as a whole new international liquidity cannot be created since the gain in transactions liquidity by one is directly offset by the surrender of it by another. Moreover, since officially held stocks of gold exceed the annual industrial demand flows by some 25 times, it follows that any substantial sales to the private sector would rapidly drive the price back down to former levels. However, there are a number of ways in which liquidity could increase, on the base provided by the increased value of gold, without depressing the price. For example, countries may use the increased value of their gold reserves to cover a rundown in their foreign currency reserves. This was precisely what the Canadian authorities achieved at the beginning of 1980 when they sold 250,000 ounces of gold at an average price of $690 in order to help finance their balance of payments deficit. Alternatively, the increase in the value of gold reserves may be used as collateral to raise a country's borrowing power or improve the terms of that borrowing.

The IMF holds some 120 million ounces of gold after completion of its gold sales programme, and these at early 1980 prices were worth some

Table 4.6. Official Gold Holdings of Central
Banks at the end of 1979.
(excluding China, USSR and associates).

	Millions of ounces	Value in $ bns*	% of Total
USA	269	188	29
West Germany	95	66	10
Switzerland	83	58	9
France	82	57	9
Italy	67	47	7
Netherlands	44	31	5
Belgium	34	24	4
Japan	24	17	3
UK	23	16	2
Canada	22	15	2
Austria	22	15	2
Portugal	22	15	2
Spain	15	11	2
South Africa	10	7	1
Venezuela	11	8	1
Lebanon	9	6	1
Australia	8	6	1
Korea	8	6	1
India	8	6	1
Algeria	6	4	
Sweden	6	4	
Saudi Arabia	5	4	
Turkey	4	3	
Argentina	4	3	
Greece	4	3	
Uruguay	4	3	
Thailand	2	1	
China Republic	2	1	
Malaysia	2	1	
Colombia	2	1	
Egypt	2	1	
Total in all central banks	930	650	100

*at $700 per ounce.
Source: Derived from *International Financial Statistics*.

$80 billion. This gold no longer has any specific role within the IMF and the Fund is free to dispose of it in any way agreed by the membership as a whole. Broad possibilities open to the IMF are:

(a) to dispose of the gold for the benefit of the developing countries (perhaps through a renewed Trust Fund arrangement using the gold

Table 4.7. Presentation of Data on Gold: Some examples.

	1972	1973	1974	1975	1976	1977
Gold, All Countries (million ounces, end of period)						
	1,017.4	1,017.5	1,016.0	1,014.8	1,009.9	1,011.7
Gold Holdings at Market Prices (million SDRs, end of period)						
All Countries	60,819	94,673	154,758	121,579	117,132	137,388
IMF	9,171	14,277	23,367	18,381	17,341	17,867
EMCF	. . . . .	. . . . .	. . . . .	. . . . .	. . . . .	. . . . .
IMF Gold Transactions (million SDRs, in period)						
Distribution (at SDR 35 per ounce)	—	—	—	—.	—	416
Sale (at SDR 35 per ounce)	—	—	—	—	137	211
Sale (at auction price)	—	—	—	—	414	761
Gold Prices and SDR Rates (end of period)						
US dollars per ounce (London)	64.90	112.25	186.50	140.25	134.75	164.95
US dollars per SDR	1.0857	1.2064	1.2244	1.1707	1.1618	1.2147
SDRs per ounce	59.78	93.05	152.33	119.80	115.98	135.79

Data: IMF, *International Financial Statistics* (August 1979).

sales to subsidise soft terms lending to the Less Developed Countries (LDCs)).

(b) to use the gold to provide prudential backing for the proposed Substitution Account.

(c) to return it to members at the original contribution rate of SDR 35 per ounce.

(d) to retain the gold in the IMF for future general purposes.

(e) to sell it to increase the Fund's liquidity for general lending.

Western central banks have been pressing for the gold to be used in backing the proposed Substitution Account. However, the LDCs are anxious that the gold should be used in their favour. The basic idea of the Substitution Account is straightforward. The account would issue claims denominated in SDRs in exchange for US dollars which could then be re-invested in US Treasury securities. Thus other countries would be able to diversify out of US dollars without creating either pressure on dollar exchange rates or a funding problem for the US Treasury. Voluntary participation has been recommended by the IMF but a wide spread of participants including developing countries would be desirable and, in order to make a useful contribution to international stabilisation, its size would have to be significant relative to existing dollar reserves (about SDR 150 billion in June 1979). The IMF study group, which prepared the proposals, has suggested that an account of SDR 50 billion would make an

Table 4.7. *(continuation)*.

	1978					1979		
I	II	III	IV	Feb	Mar	Apr	May	June
1,013.1	1,013.2	1,013.9	1,018.6	1,015.8	933.8	932.2	930.5	. . . .
148,773	149,629	171,819	176,692	198,006	174,227	179,683	201 371	. . . . .
18,926	18,789	21,280	20,505	22,774	21,702	22,330	24 961	24 698
. . . . . .	. . . . .	. . . . .	. . . . .	—	15,026	15 522	17,427	. . . . .
3	9	—	201	—	2	—	—	—
92	49	57	58	19	17	17	17	16
226	356	258	271	103	88	89	95	96
181.60	183.05	217.10	226.00	251.30	240.10	245.30	274.60	277.50
1.2367	1.2395	1.2811	1.3028	1.2892	1.2868	1.2727	1.2689	1.2911
146.00	7.68	169.47	173.47	194.93	186.58	192.74	216.40	214.93

important contribution, as not all dollar holdings are subject to diversification pressure.

The main difficulties in achieving agreement on the Account have concerned the yield and liquidity of countries' SDR holdings and the provisions for covering any running or capital losses on the Account. The Germans frequently argue that the substitution scheme is likely to be inflationary not through its direct monetary effects, but by relaxing some of the pressures, which at present induce some governments, especially the USA to curb domestic credit expansion. Pressures from the French and Germans have resulted in the substitution account being placed on ice (*see Chapter 8*). Use of the IMF gold windfall to aid LDCs is more likely to lead to direct increases in world demand since the recipients would receive the aid as an inflow of hard currencies and would be likely to increase their imports in consequence.

All members of the European Monetary System, including the UK, have deposited 20 per cent of their gold (and 20 per cent of their dollar) reserves with the European Monetary Corporation Fund and have received European Currency Units (ECUs) in return. (*See Appendix II of Chapter 3.*) These ECUs are liquid and may be used to settle intervention debts between members of the exchange rate mechanism of the EMS. The number of ECUs issued to members depends on the average price of gold over the previous six months. It follows that rises in the gold price, if sustained, will lead to a direct increase in the liquid reserves available to help defend currencies at their lower intervention limit. This arrangement can be of particular benefit to large gold holders in the Community (notably France and Italy) because it will allow them to effectively cash in some of their gold at high price levels without precipitating the price reduction likely to follow direct gold sales. The extent to which the EMS arrangements increase European money supply seems however to be fairly strictly limited by the detailed intervention regulations.

Revised IMF Presentation Data on Gold

In line with the increased importance of gold, the IMF, with effect from August 1979, announced a revised presentation of its international reserve statistics as published in *International Financial Statistics (see Table 4.7)*. A new format for world tables provides data on individual countries' gold holdings in physical quantities and measures the value of official holdings for all countries, and the world, both at SDR 35 per troy ounce of fine gold, the former official price, and at prevailing market prices. The new presentation attempts to reflect current attitudes toward gold as a measure of international liquidity, particularly its valuation, while continuing to report the gold component of reserves in an internationally comparable form.

The revised format for the world tables consists of the customary tables on 'SDRs', 'Reserve Positions in the Fund', and 'Foreign Exchange'; the sum of these three components is reported in a new table on 'Total Reserves minus Gold'; the usual table on 'Total Reserves' (with gold valued at SDR 35 per ounce); and a new table on 'Gold' expressed in millions of troy ounces replaces the table on 'Gold' valued at SDR 35 per ounce.

The data on gold in physical quantities are supplemented by foot-tables to the main table showing gold holdings (for the categories All Countries, IMF, European Monetary Cooperation Fund (EMCF), and the World) valued at SDR 35 per ounce and at market prices, IMF gold transactions, and the London market price for gold in US dollars, the US dollar per SDR rate, and the derived market price for gold in SDRs. The London market price for gold is also given as a separate line on the country page for the United Kingdom. The 'Total Reserves' table is supplemented by foot-tables on total reserves, with gold valued both at SDR 35 per ounce and at market prices for the categories All Countries, IMF, EMCF, and the World. The foot-table data on gold are not given for individual countries.

The IFS country pages now include two new standard series on each Fund member's total reserves minus gold as well as each one's official gold holdings in physical terms. In addition, a new line, for the US dollar value of gold in accordance with national valuation practices, is added to the country pages.

For each country, the valuation of gold included in the monetary authorities' data on foreign assets, as published in the money and banking sections of IFS, will continue to follow national practices. These are described in the notes to the country pages of IFS.

References

(1) Williamson, J. International Liquidity: A Survey, *Economic Journal*, September 1973.
(2) Sorensen, T. C. *Kennedy*. Harper and Row: New York 1973.
(3) Triffin, R. *Gold and the Dollar Crisis: The Future of Convertibility*. Revised edition. Yale University Press: New Haven 1961.
(4) Op. cit.
(5) Rueff, J. *Balance of Payments Proposals for the Resolution of the most Pressing World Economic Problem of Our Time* (Macmillan: New York 1967).
(6) Bergsten, C. F. *The Dilemmas of the Dollar*. New York University Press 1975.
(7) Williamson, J. *The Failure of World Monetary Reform 1971-1974* (Thomas Nelson 1977).
(8) Rees-Mogg, W. *The Reigning Error: the Crisis of World Inflation*. Hamilton: London, 1974.
(9) Vilar, P. *A History of Gold and Money 1450-1920*. NLB London 1976.

Chapter Five

GOLD: THE MAJOR FACTORS AFFECTING ITS SUPPLY, DEMAND AND PRICE

Introduction

As discussed in Chapter 3, prior to 1976 the gold price was subject to extensive intervention by central banks with the result that the degree of price fluctuation was severely limited. Increasing difficulties, particularly, after 1971 made this price intervention almost impossible to achieve. For the various reasons outlined in Chapter 3 the International Monetary Fund took the decision, in 1976, to abolish the official gold price, a price which had by then, become purely notional. This decision, which effectively allows the gold price to be determined by the interaction of supply and demand, has focussed attention on the factors affecting both the supply of and demand for gold. In this chapter we examine the basic factors affecting both the supplies of newly mined gold coming onto the markets and illustrate who purchases all this new gold. The final section will put this analysis together to illustrate the principal factors affecting the gold price. First of all let us examine how markets operate.

Fundamentals of Market Analysis

The supply of an imperishable storeable commodity consists of new production and reductions of stock and the demand for it is composed of current consumption and increases in stocks. Gold is an eminently storeable commodity with stocks of a size many times annual output.

The demand for gold can be analysed in terms of income and substitution effects. Generally, as incomes and affluence increase from year to year the demand, for gold jewellery in particular, increases (with other factors held constant). The effects of incomes upon the industrial demand for gold are less marked with this demand being determined more strongly by technological changes and innovations than by income effects. Lipschitz and Otani have estimated an income elasticity of demand for gold at 0.6 (i.e. 1 per cent increase in incomes results in a 0.6 per cent increase in the demand for gold).[1] These results are however completely altered by the rising importance of the Organisation of Petrol Exporting Countries (OPEC).

The substitution of gold for competitive materials such as silver in jewellery manufacture is determined to a large extent by gold's decisive lead in prestige as jewellery over silver and other materials rather than its physical characteristics. Thus if the silver price falls consumers are unlikely to substitute silver jewellery for gold jewellery. However, if the price of gold itself changes, a different picture emerges. With a fall in the price of gold, this may directly reduce the demand for gold as it becomes a less prestigious product. However, if the price of gold were to drop sufficiently, gold would become competitive again in its own right owing to its physical properties, described in Chapter 1, rather than to its prestige, and the quantity of gold demanded would rise again. Lipschitz and Otani estimate the price elasticity of demand for gold at 0.7 (i.e. a 1 per cent rise (fall) in the price of gold results in a 0.7 per cent fall (rise) in the demand for gold). Again these results are altered by the existence of OPEC. J. Aron Commodities Corporation have suggested that the prestige of gold is such that a rise in the price of gold results in an increase in the demand for gold. This they have termed a 'perverse demand elasticity'.

The relationship between a change in the gold price and new gold coming onto the market is not simple. In the long run, low prices mean low output and vice versa; but it takes about seven years to develop a new mine, so there can be a long time-lag. Moreover, a rise in price reduces the minimum gold content of ore that can be processed to yield a profit, and so may lead mines to concentrate on lower grade ores, keeping higher grade ores for the future; in South Africa this is encouraged by a progressive profits tax intended to prolong the life of the mines. Supply is, of course, also influenced by the discovery of new sources and by improvements in technology which are partly independent of price, though a high price encourages exploration and research in the long run. Lipschitz and Otani estimated the price elasticity of the supply of gold to be 0.1 (i.e. a 1 per cent increase (decrease) in the price of gold results in a 0.1 per cent increase (decrease) in the supply of gold.[2]

The predictive value of econometric analysis such as that undertaken by Lipschitz and Otani is reduced, however, by the poor quality of statistics on available above-ground stocks. In addition, there is the problem that higher gold prices, which would stimulate output and thus increase the avavilable supply in some countries, might lead holders of large stocks, such as South Africa and the Soviet Union, to reduce sales. Each of the components of supply and demand may be subject to change: production may change because of technological factors, consumption because of income and substitution effects, private stocks because of changing expectations, and monetary stocks because of changing attitudes of central banks. The various components of the supply and demand for gold are summarised in Table 5.1.

Table 5.1. The Components of the Supply and Demand for Gold

Demand	Supply
Current consumption	Current production
— in jewellery	Reprocessed scrap
— in dentistry	Reductions of processors' inventories
— in industry	Reductions of other private stocks
Additions to processors' inventories	Reductions of monetary stocks
Additions to private stocks	
Additions to monetary stocks	

The Demand for Gold

The most comprehensive source of information with regard to gold supply and demand on a worldwide basis is a series of annual reports published by Consolidated Gold Fields Limited. The majority of statistics in this chapter are taken from their publication *Gold 1981*. This breaks down the private demand for new gold into the combined requirements for fabrication and investment purposes. The fabrication demand is defined as the transformation of gold in bar form to semi-manufactured or final product for industrial or commercial use. Thus the fabrication demand for gold consists of gold for jewellery manufacture, for electronic and other industrial applications, for dentistry, for official gold coins and for medals, medallions and fake coins. While purchases for these purposes are susceptible to changes in the business cycle, and to changes in the price of gold like purchases of other commodities, they can be estimated with some degree of confidence. On the other hand, purchases of gold for official or monetary purposes or for private investment or hoarding are subject to unpredictable variations, which arise in considerable measure from political, economic or financial changes, and the changes in interest rates and inflation rates which accompany them.

Although the purchase of jewellery can be regarded as investment it can still be useful to break down the demand for gold into the following components:

(1) Consumption of gold
    (a) Jewellery and art
    (b) Dentistry
    (c) Industry
(2) Stockpiling of gold
    (a) Private
        (i) Short-term speculation
        (ii) Long-term speculation
        (iii) Hoarding for safekeeping
        (iv) Hoarding for traditional reasons
    (b) Official: for monetary reserves

Table 5.2 gives the overall demand for new gold in the years 1970 to 1980 for both developed and developing countries. A fabrication figure for 1980 is given as 521 tonnes. To this Consolidated Gold Fields Limited adds net private bullion purchases in order to derive the non-communist private sector demand for gold. Net private bullion purchases not always being observable directly are determined by subtracting the use of gold in product fabrication from the supplies to the non-communist private sector.

Carat Jewellery and the Arts

Pure gold is too soft and liable to wear to be used extensively in jewellery and, from ancient times therefore, it has always been alloyed with other metals. The purity of gold alloy is expressed in terms of carats. The word 'carat' derives from the Italian 'carato', the Arabic 'qirat' or the Greek 'keration'. These all mean 'the fruit of the carob tree'. This tre, has horn-like pods containing seeds which were once used to balance the scales in Oriental bazaars. The carat number indicates the relative purity of an alloy. Pure gold is described as having 24 carats; a piece of jewellery in 18 carat gold is made of an alloy containing 18 parts gold out of 24. The rest is another metal, usually copper or silver. Fineness is a metallurgical term indicating the purity of the metal. For example, gold that has a millesimal fineness of 916.6 is 916.6 parts out of 1,000 gold. Thus, as mentioned 24 carat gold is fine gold and any other carat number is the amount of fine gold as a fraction of 24.

It is the jewellery at the lower end of the carat scale that provides the greatest demand for gold. In Britain, 80 per cent of gold trade is in 9 carat gold. In the USA, Handy and Harman, the major suppliers of carat gold alloys to the jewellery industry, report that the largest slice of gold goes to 10 carat alloys for the millions of school award pins, fraternity pins and emblems made each year. The present legal standards for gold in Britain are:

22 carat – 916.6 parts gold in 1000 (22 out of 24)
18 carat – 750 parts gold in 1000 (18 out of 24)
14 carat – 585 parts gold in 1000 (14 out of 24)
 9 carat – 375 parts gold in 1000 (9 out of 24)

Gold and its alloys are used in carat jewellery and the arts because of their beauty, their resistance to tarnish, their relative ease of working and their intrinsic value. This does not mean that economic and technical considerations are not highly significant in dictating the manner, form and caratage in which gold jewellery is fabricated. Nevertheless consumer level traditional social customs and fashion trends have strong influences on the market for gold jewellery.

Typical of fashion trends was the swing away from the use of gold in jewellery in favour of silver and 'white alloys' which coincided with the

Table 5.2. The Use of Gold by the Non-Communist Private Sector.
(Metric tons)

	1970	1971	1972	1973	1974	1975	1976	1977	1978	1979	1980
A Total increase in gold holdings	1034	1386	1244	1398	1246	1112	1440	1642	1751	1704	803
Fabricated gold in developed countries											
Carat jewellery	500	553	702	428	278	317	471	540	592	551	270
Electronics	89	86	105	126	91	66	75	76	84	92	79
Dentistry	55	59	61	64	54	58	73	78	85	82	60
Other industrial/ decorative uses	58	62	65	67	64	55	59	59	70	69	61
Medals, medallions and fake coins	29	33	32	19	11	10	20	22	21	16	18
Official coins	32	44	44	36	209	221	145	125	256	242	164
B Total	763	837	1009	740	707	727	843	900	1108	1052	652

Table 5.2. (continuation).

	1970	1971	1972	1973	1974	1975	1976	1977	1978	1979	1980
Fabricated gold in developing countries											
Carat jewellery	566	511	297	90	−53	206	464	463	415	180	−150
Electronics	4	4	5	1	1	1	1	1	2	2	2
Dentistry	4	6	5	4	3	4	4	4	4	4	2
Other industrial/ decorative uses	4	6	5	4	3	4	4	5	7	6	3
Medals, medallions and fake coins	25	19	9	2	−4	11	30	29	29	17	−3
Official coins	14	10	19	18	78	30	37	17	31	48	15
C Total	613	550	335	119	28	256	540	519	488	263	−131
D Total fabricated gold (B + C)	1376	1387	1344	859	735	983	1383	1419	1596	1315	521
E Bullion holdings* (A − D)	−342	−1	−100	539	511	129	57	223	155	389	282

*This category excludes coins, but includes hoarding of small bars and all other forms of bullion investment.
Source: *Gold 1981* published by Consolidated Gold Fields Limited.

Table 5.3. Fabrication of gold in carat jewellery (net).
(Metric tons)

	1970	1971	1972	1973	1974	1975	1976	1977	1978	1979	1980
Europe											
Italy	160.5	174.3	313.0*	98.2	50.0	76.0	177.0	209.0	235.0	227.0	87.0
Spain	53.0	55.0	58.0	45.2	22.5	37.8	45.5	49.7	51.5	37.7	11.4
Germany	47.6	47.1	45.4	38.0	25.0	27.0	36.0	44.0	47.0	45.5	28.5
UK & Ireland	14.7	15.5	18.0	20.3	15.8	18.3	19.6	24.0	22.0	21.7	9.1
France	25.7	30.8	29.4	22.4	14.6	17.9	24.5	26.2	25.3	24.4	11.7
Switzerland	23.0	23.0	22.0	16.0	13.5	8.2	10.9	15.5	17.7	15.3	12.2
Greece	11.0	12.0	13.5	13.5	7.0	9.5	8.3	11.0	14.0	10.0	4.0
Belgium	4.2	4.8	5.5	6.5	5.5	6.0	6.6	7.4	6.1	5.0	1.5
Portugal	11.0	13.9	12.3	8.9	4.4	2.9	5.8	5.5	4.5	4.0	1.3
Yugoslavia	—	—	—	0.3	3.0	3.0	4.0	4.5	4.5	4.5	5.9
Netherlands	2.9	3.4	3.4	2.4	1.4	1.7	2.4	3.0	2.2	1.8	0.6
Austria	6.1	6.8	6.6	4.1	2.5	2.0	3.5	3.6	2.9	2.6	0.7
Sweden	2.0	2.7	2.7	2.3	1.8	1.8	1.8	2.1	1.7	1.3	0.8
Denmark	3.6	3.1	4.1	2.6	0.9	0.9	1.7	2.0	1.4	1.1	0.3
Finland	2.2	1.9	2.0	1.8	1.2	1.1	1.4	1.6	1.6	1.4	0.6
Cyprus & Malta	0.8	0.8	0.6	0.6	0.2	0.2	0.7	1.0	1.3	1.1	0.3
Norway	1.3	1.3	1.0	0.5	0.5	0.5	0.8	0.8	0.7	0.8	0.3
Total Europe	369.6	396.4	537.5	283.6	169.8	214.8	350.5	410.9	439.4	405.2	176.2
North America											
USA	97.8	125.8	127.3	102.0	61.1	58.7	72.8	79.5	83.4	79.5	57.8
Canada	5.9	6.7	9.0	9.7	9.4	9.2	12.2	17.7	20.3	17.0	7.0
Total North America	103.7	132.5	136.3	111.7	70.5	67.9	85.0	97.2	103.7	96.5	64.8

Table 5.3. *(continued).*

	1970	1971	1972	1973	1974	1975	1976	1977	1978	1979	1980
Latin America											
Brazil	29.0	30.0	19.0	12.0	10.0	13.0	16.2	20.4	29.0	33.0	14.0
Mexico	21.0	20.0	15.5	7.8	5.0	6.1	10.5	7.9	9.3	7.4	3.0
Venezuela	5.0	4.9	4.2	0.3	−1.0	0.3	1.5	1.8	3.4	2.8	1.3
Colombia	2.8	2.9	1.7	1.9	0.8	1.2	1.0	2.0	2.0	1.9	0.4
Peru	4.2	3.7	3.0	1.4	1.3	1.2	1.2	0.8	0.8	0.5	−4.0
Chile	0.6	1.3	0.8	0.8	—	—	—	0.1	0.4	0.3	—
Argentina	16.5	12.0	0.5	—	−2.0	−10.0	−10.0	−8.0	−5.0	−2.0	−4.0
Other Latin America	2.2	2.1	2.0	⎰ 1.0							
Central America & Caribbean	3.5	4.3	2.2	⎱	−0.5	0.5	1.5	4.9	5.2	3.7	−3.0
Total Latin America	84.8	81.2	48.9	25.2	13.6	12.3	21.9	29.9	45.1	47.6	7.7
Middle East											
Turkey	32.0	23.0	13.1	3.0	12.0	41.4	100.7	80.0	86.0	10.0	−18.0
Saudi Arabia & Yemen	2.5	2.5	2.0	−2.0	−1.0	8.0	33.0	34.5	30.0	7.5	2.0
Iran	13.0	14.0	11.0	7.0	5.0	25.0	53.0	64.0	30.0	−4.0	−70.0
Iraq, Syria & Jordan	7.5	7.0	3.5	−18.0	−23.0	10.0	15.0	17.0	21.1	6.1	−5.0
Egypt	6.5	7.0	3.5	−6.0	−23.0	6.0	25.5	21.0	20.5	12.5	−5.0
Arabian Gulf States	5.0	9.1	4.1	−12.0	−9.0	9.0	⎰ 13.0	17.2	16.0	13.7	2.6
Kuwait				3.0			⎱ 9.0	14.5	13.3	9.0	1.0
Israel	5.0	5.0	4.0	3.0	2.9	2.9	4.0	5.6	7.4	7.5	4.0
Lebanon	3.0	4.0	3.0	−2.0	1.0	2.0	2.0	2.0	2.0	3.0	−1.0
Total Middle East	74.5	71.6	44.2	−27.0	−35.1	104.3	255.2	255.8	226.3	65.3	−89.4

Table 5.3. (continued)

	1970	1971	1972	1973	1974	1975	1976	1977	1978	1979	1980
Indian Sub-Continent											
India	215.0	175.0	107.2	60.5	14.0	25.0	32.9	39.5	44.0	10.0	−9.0
Pakistan & Afghanistan	30.0	25.0	15.0	4.0	−0.5	5.5	20.0	15.0	5.0	2.0	−2.5
Sri Lanka	4.0	6.0	3.0	2.0	−1.0	−0.5	2.0	1.5	0.5	0.3	−0.5
Bangladesh & Nepal	—	—	−3.0	−3.0	−6.5	−3.5	−2.0	1.0	3.0	0.3	−1.0
Total Indian Sub-Continent	249.0	206.0	122.2	63.5	6.0	26.5	52.9	57.0	52.5	12.6	−13.0
Far East											
Hong Kong	10.0	11.0	13.0	10.0	−1.0	0.5	7.5	8.0	11.0	14.0	3.0
Thailand	12.0	8.0	−10.0	−24.0	−15.0	5.0	12.0	10.0	10.0	5.0	−2.0
Singapore	5.0	5.0	4.2	4.2	0.5	3.0	6.3	8.5	9.0	7.0	−1.0
Taiwan	4.0	4.5	2.0	5.0	—	6.0	6.5	5.6	7.0	5.0	2.0
Malaysia	6.5	6.0	5.0	2.0	0.3	4.0	7.1	6.0	6.8	5.5	0.5
Philippines	5.0	6.0	6.0	2.0	−3.0	0.8	2.5	3.7	4.0	2.8	0.5
South Korea	6.0	6.0	4.0	3.0	—	1.5	2.0	2.0	3.0	2.5	1.0
Indonesia	30.0	25.0	10.0	2.5	−15.0	15.0	35.0	30.0	−3.0	−8.0	−55.0
Vietnam	10.0	8.0	3.5	−6.0	−5.0	−5.0	−0.2	} −0.5	—	−3.0	−8.0**
Burma, Laos & Kampuchea	7.0	6.0	−1.0	−3.0	−14.0	−2.0	−0.1				
Total excluding Japan	95.5	85.5	36.7	−4.3	−52.2	28.8	78.6	73.3	47.8	30.8	−59.0
Japan	34.5	34.5	40.0	45.0	39.9	38.6	40.1	38.5	56.7	53.5	26.3
Total Far East	130.0	120.0	76.7	40.7	−12.3	67.4	118.7	111.8	104.5	84.3	−32.7

Table 5.3. (continued).

	1970	1971	1972	1973	1974	1975	1976	1977	1978	1979	1980
Africa											
Morocco	23.0	25.0	12.0	9.0	5.4	12.0	25.0	22.0	21.7	16.0	—
Algeria	3.0	5.0	2.5	1.5	1.0	1.5	2.0	2.0	1.8	1.2	—
Libya	7.5	4.2	2.0	1.4	1.2	3.7	5.6	4.5	4.0	1.4	1.7
South Africa	3.2	3.5	3.2	3.0	2.8	1.4	2.0	2.0	2.0	1.4	1.0
Other Africa	9.0	9.0	5.0	1.0	-2.0	6.0	10.0	5.0	2.5	1.7	—
Tunisia	2.0	2.0	1.5	—	—	1.0	1.5	1.5	1.0	1.7	0.2
Total Africa	47.7	48.7	26.2	15.9	8.4	25.6	46.1	37.0	33.0	23.4	2.9
Australia	6.9	7.1	7.1	4.7	4.1	3.7	4.4	3.0	2.7	2.1	3.0
Total non-communist world	1066.2	1063.5	999.1	518.3	225.0	522.5	934.7	1002.6	1007.2	737.0	119.5

*Includes inventories built up before introduction of VAT
**Revenues of large inflows of gold which took place prior to the change in political classification of Vietnam, Laos and Kampuchea.
Source: Gold 1981 published by Consolidated Gold Fields Limited.

rapid increase in the price of gold in 1972 and 1973. The 1974 rise in the gold price resulted in a further decline in the demand for carat jewellery. In 1974 demand for gold jewellery dropped to 7.5 million ounces. 1975 and 1976 were characterised by a rise in the demand for jewellery, with the situation levelling off in 1977 and 1978. Consolidated Gold Fields Limited give an estimated demand for new jewellery in 1980 of 119.5 tonnes, well down on the earlier years of the 1970s (see Table 5.3).

Dentistry and Medical Uses

The responsiveness of the quantity of gold demanded for utilisation in dentistry to changes in the price of gold is difficult to assess. As in the case of jewellery, the gold content in the final price to the consumer is relatively low, with the labour content more significant. One must ask the question why gold is preferred over other substitutes as a filling material for teeth? If gold is considered to be superior because of its physical properties, such as a more permanent filling or as causing less discoloration to filled teeth, then it should be relatively unaffected by a rise in its price. Gold may also be preferred because gold filled teeth have a prestige advantage over say, completely white teeth that are filled with ceramic, which is a cheaper process. Since both these factors are likely to have an effect, the demand for gold for dentistry is likely to be somewhat price insensitive. The dentistry demand is also influenced by the existence of state health insurance schemes.

Details of the demand for gold in the manufacture of dental alloys taken from *Gold 1981* are given in Table 5.4. The demand is dominated by West Germany, USA and Japan. Overall the demand has remained stable between 1977 and 1979 despite variations in the gold price although it did fall significantly in Japan and the USA in 1980. The use of gold in dental applications in West Germany has soared since 1975 when new social security coverage made it possible for gold work to be covered by insurance. Initially 100 per cent of gold dental work was covered by insurance, but this has now been reduced to 80 per cent.

Quite apart from its long standing dental application, gold does have a limited number of medical uses. It has been used in Europe since 1927 for the treatment of rheumatoid arthritis. It is administered intra-muscularly as a soluble salt in cautiously increased doses to a level of 25 mg. per week. It is a slow treatment, with some benefit after six weeks, but maximum benefit after six months. Gold has also been used tentatively in the treatment of cancer – in the form of an injection of a colloidal suspension of radioactive gold. For entirely different reasons gold is used in X-ray machines as the target which arrests the electron beam. Solid gold barriers can be used internally in some patients to protect their vital organs when X-ray or radiation treatment is carried out.

Electronics

Some of gold's properties, which were discussed in Chapter 1, give it

Table 5.4. Gold Use in Dentistry.
(Metric tons)

	1970	1971	1972	1973	1974	1975	1976	1977	1978	1979	1980
Germany	8.5	9.0	11.0	12.0	10.0	12.0	20.1	24.7	27.0	28.0	25.2
United States	20.4	23.3	23.3	21.1	16.5	19.9	21.6	23.6	24.0	21.2	13.8
Japan	9.6	10.0	10.3	12.5	10.0	10.0	13.0	10.5	13.3	12.0	5.9
Italy	4.0	4.5	4.0	4.0	4.0	3.0	3.5	4.0	4.5	3.6	2.5
Switzerland	1.3	1.3	1.4	4.0	4.0	3.8	3.9	4.2	4.4	4.9	4.8
France	2.5	2.7	3.0	3.0	2.5	2.5	2.8	3.0	3.4	3.4	2.0
South Africa	—	—	0.2	0.1	0.1	0.1	0.1	1.3	1.4	1.8	1.0
Israel	—	—	—	1.0	1.3	1.3	1.3	0.8	1.2	1.5	0.8
Brazil	1.0	1.0	1.0	1.0	1.0	1.1	1.1	1.1	1.1	1.1	0.5
Yugoslavia	—	—	—	—	0.5	0.5	0.8	1.1	1.1	1.1	0.9
Mexico	1.3	1.7	2.1	1.8	0.5	1.0	1.0	1.0	1.0	1.0	0.3
Australia	0.5	0.6	0.6	0.6	0.5	0.4	0.8	0.8	0.7	0.5	0.3
Netherlands	0.9	0.9	1.1	0.9	0.8	0.8	1.0	0.8	0.8	0.8	0.7
Sweden	1.2	0.9	0.8	0.7	1.3	1.3	1.0	1.0	0.8	0.6	0.4
United Kingdom	1.0	0.8	0.8	0.7	0.6	0.8	0.8	0.8	0.8	0.8	0.6
Austria	0.9	1.1	1.2	0.9	0.8	0.7	0.7	0.7	0.7	0.7	0.3
Greece	1.3	1.0	1.0	0.9	0.8	1.0	0.7	0.5	0.5	0.6	0.2
Belgium	1.2	1.3	1.3	0.3	0.4	0.4	0.4	0.4	0.4	0.4	0.4
Spain	1.2	1.2	1.3	0.8	0.5	0.4	0.5	0.4	0.4	0.4	0.2
South Korea	—	—	—	—	—	0.4	0.6	0.8	0.9	1.0	0.3
Canada	0.7	0.6	0.5	0.5	0.3	0.3	0.3	0.3	0.2	0.2	0.1
Denmark	0.1	0.1	0.1	0.1	0.1	0.1	0.1	0.2	0.2	0.1	0.1
Norway	0.2	0.2	0.2	0.2	0.1	0.1	0.1	0.1	0.1	0.1	0.1
Peru	0.2	0.2	0.2	0.1	0.1	0.1	0.1	0.1	0.1	0.1	0.1
Portugal	0.1	0.1	0.1	0.1	0.1	0.1	0.1	0.1	0.1	0.1	0.1
Venezuela	0.4	0.4	0.3	0.2	0.1	0.1	0.1	0.1	0.1	0.1	0.1
Total	58.5	62.9	65.8	67.5	56.9	62.2	76.5	82.4	89.2	86.1	61.7

Source: *Gold 1981* published by Consolidated Gold Fields Limited.

special importance for the electronics industry. Gold is an exceedingly good conductor of electricity and heat. Only silver is a better conductor of electricity, but it does not have the same resistance to tarnish and corrosion as gold. In addition, when two gold surfaces are brought into contact with one another, electrical connection between them is readily established, and the electrical resistance over the contact area, i.e. the contact resistance, is low and stable. Other metals cannot match gold in this respect.

Gold is exceptional in regard to the ease with which it can be bonded to other metals. One reason for this is its freedom from obstructive tarnish films on its surface. As a result, it is easily soldered even after long exposure to corrosive atmospheres and can even be bonded to other metals without the use of solders, by the simple application of pressure or of heat plus pressure. Again, other metals cannot compete with gold in this respect. Gold is a very efficient reflector of heat rays, and since it does not tarnish, it retains its reflectivity even after long exposure in atmospheres which cause other metals to corrode or tarnish. Not only does gold mix readily with a variety of other metals to form alloys, but these alloys often themselves possess exceptional properties.

In the light of its special properties listed above, it will readily be appreciated why gold is such as important metal in the electrical, and more particularly, the electronics and communications industries. In the latter, the voltages are so small, the circuitry so complex, and the reliability which is called for so high, that the metals used for conducting elements and contacts must meet the most stringent requirements in respect of electrical conductivity, contact resistance, bondability and resistance to the environment. Gold has a unique combination of such properties and is therefore extensively used in the components from which computers, calculators, telephone systems, radio and TV equipment, the control systems of missiles and spacecraft and much of the equipment which is so commonly used for the control and automation of industrial processes are made. In such products gold is most extensively used in electrical contacts and connectors. In some instances the contacting areas of these are coated with a thin film of gold or gold alloy, while in others, small pieces of solid gold or gold alloy, are welded to them. Although the mass of gold employed in any one contact or connector may be very small indeed, the number used in telephone systems, computers and electronic devices generally is astronomically large. The total use of gold in electrical contacts and connectors is therefore considerable.

Gold is also used extensively for the electrically conducting elements in electronic circuitry. Thus, very fine gold wire is used to join the units in such circuitry in some situations, while in others the active components are linked by thin gold coatings in appropriate patterns of lines upon a non-conducting ceramic base. The coatings may be applied by electro-deposition, by vacuum techniques such as evaporation, sputtering, or what is known as 'thick film technology'. In this latter process the circuitry

Table 5.5. Gold Use in Electronics.
(Metric tons)

	1970	1971	1972	1973	1974	1975	1976	1977	1978	1979	1980
United States	44.8	42.8	51.2	61.6	44.0	24.9	27.7	31.4	32.4	34.7	30.7
Japan	18.1	20.1	22.5	26.0	15.0	17.1	23.1	19.5	25.0	25.6	22.7
Germany	10.0	8.0	11.0	15.0	12.0	8.0	8.2	8.0	8.2	9.8	8.3
Switzerland	0.9	1.0	1.5	3.0	3.0	2.5	3.1	3.7	4.4	4.5	3.6
France	4.5	4.0	4.5	5.0	5.5	4.5	3.7	3.8	4.0	4.2	3.7
United Kingdom	5.8	4.0	5.0	5.5	4.4	3.0	2.7	2.8	3.0	6.0	4.8
Italy	1.5	1.6	5.0	5.0	2.0	1.9	2.3	2.4	2.9	3.1	2.4
Netherlands	1.1	1.3	1.4	1.6	1.5	1.0	1.2	1.2	1.5	1.8	0.9
Yugoslavia	—	—	—	—	1.0	1.0	1.0	1.0	1.0	1.0	0.9
Singapore	—	—	0.3	0.8	0.3	0.3	0.3	0.3	0.9	0.5	0.5
Canada	0.5	0.7	0.8	1.1	0.9	1.0	0.4	0.5	0.7	0.9	0.3
Taiwan	—	—	—	—	—	—	0.1	0.2	0.2	0.2	0.2
Spain	0.7	0.8	0.8	0.6	0.7	0.7	0.6	0.5	0.4	0.4	0.3
Austria	1.0	0.8	0.8	0.5	0.5	0.4	0.2	0.3	0.3	0.3	0.2
Brazil	—	—	—	0.2	0.2	0.2	0.3	0.3	0.3	0.4	0.6
Australia	0.4	0.6	0.5	0.6	0.5	0.5	0.3	0.2	0.2	0.1	0.1
India	—	—	0.1	0.1	0.2	0.2	0.1	0.1	0.2	0.2	0.2
South Korea	—	—	—	—	—	0.1	0.2	0.2	0.2	0.3	0.2
Mexico	—	—	—	—	—	—	—	0.1	0.1	0.1	0.1
Total	89.3	85.7	105.4	126.6	91.7	67.3	75.5	76.5	85.9	94.1	80.7

Source: *Gold 1981* published by Consolidated Gold Fields Limited.

is 'printed' onto the ceramic base using a gold-containing paste or 'ink', which leaves a residue of gold on the substrate after its non-gold constituents have been burned off. In some instances where excessive heat is generated in an electronic device, the dissipation of this heat is necessary and is assisted by 'thick' layers of gold in appropriate positions. Semiconductor elements in special microwave transistors and diodes may be protected in this way.

Another important use of gold is in the 'doping' of semi-conductors, especially those which are used in high-speed computing devices. The introduction of carefully controlled quantities of gold into, for example, silicon semi-conductors improves their working properties, and the use of such 'doped' semi-conductors has contributed significantly to the development of modern computing equipment.

In industry, economic considerations dictate, first, that gold be used only where its special properties cannot be matched by those of cheaper metals or materials and second, that it is used with the utmost economy. In industrial applications therefore it is used mostly in the form of fine wires or of thin coatings or films, and where feasible it is alloyed with a cheaper metal.

As can be seen from Table 5.5 gold's use in the manufacture of electronic components increased from 86 tonnes in 1978 to 94 tonnes in 1979 but fell to 81 tonnes in 1980. This is well below the figure of 126 tonnes reached in 1973. As might be expected, the demand for gold for the fabrication of electronic equipment is confined largely to developed countries with the USA, Japan and West Germany consistently being the largest consumers.

The dominant users of gold in electronics, as can be seen from Table 5.5, are the USA and Japan.

Other Industrial and Decorative Uses

Within this sector *Gold 1981* includes gold uses in industrial and decorative plating (excluding electronics), liquid gold used in ceramics, and rolled gold (gold fill). According to Consolidated Gold Fields Limited (*see Table 5.2*) there was a reduction to 64 tonnes in 1980 compared with 75 tonnes in 1979. The USA is the major user in the world and the other two countries which make a major contribution to this sector are Japan and West Germany.

Gold-plated glass is currently being promoted as both an energy saver and a means of improving the appearance of buildings. It forms a window which permits visible light to come through but stops the heat waves. Air conditioning costs in summer can also be reduced because the infrared heat waves are reflected back into the outside air. During the winter the window reflects indoor heat back into the building and saves energy in the reverse manner. Additional savings are made in capital costs because a building which is fitted with gold-plated glass requires smaller capacity equipment for heating, ventilating and air-conditioning.

Medals, Medallions and Fake Coins

Fake coin manufacture is a traditional industry in Saudi Arabia and Kuwait and the designs range from copies of ancient Turkish coins right up to modern coinage, including the British sovereign. As one would expect, quality of reproduction varies widely, but the gold content is usually up to the standard of the official version. The Iranian revolution resulted in Kuwait losing its best market for fake coins. In the period 1975 to 1979 the fake coin business flourished in Saudi Arabia because of the increase in purchasing power and extensive buying by immigrant workers. Table 5.2 gives the statistics for new gold used in medals, medallions and fake coins. Demand in this category declined substantially in 1980 falling to 15 tonnes. This can be compared with figures of 50 and 33 tonnes in 1978 and 1979, respectively. The major market is now the USA which in 1980 began a programme for the production of gold medallions. The USA plans to strike a series of medallions over a period of five years at a rate of 1,000,000 ounces per year.

Official Coins

According to *Gold 1981* and illustrated in Table 5.2 gold used in official coins has increased substantially. The major issuers of official gold coins are South Africa, United Kingdom, Mexico and Canada. The high levels of official coin production established in 1978 were sustained during 1979, an estimated 290 tonnes of gold being consumed compared with 288 tonnes in 1978. During 1980 an estimated 234 tonnes of gold (net) was issued in the manufacture of official coins.

The South African Krugerrand was, in 1980, again the dominant coin in world markets. The South African Krugerrand is a gold bullion coin that is legal tender in its country of origin. It contains exactly one troy ounce of fine gold and weighs 1.0909 troy ounces in all. (The troy system is used for measuring precious metals and is based on a pound of 12 ounces or 31.1 grams.) Gold of 24 carat is too soft to make a sturdy usable and desirable gold coin so the Krugerrand is alloyed with pure copper, making it 22 carat (0.91666 fine) but still containing one ounce of pure, 24 carat gold. The Maple Leaf is also 24 carat of gold.

In September 1980 the South African Chamber of Mines issued three new 'mini' Krugerrands. The new Krugerrands contain one half, one quarter and one tenth of a troy ounce of pure gold respectively. The new Krugerrands are to be marketed as bullion coins with premiums over the value of the gold content of 5 per cent for the half ounce version, 7 per cent for the quarter ounce, and 9 per cent for the one-tenth ounce model.

The Krugerrand was first minted in 1967 in limited quantities and in larger numbers from 1980. In 1976 the public bought 3,004,945 newly minted Krugerrands. In 1977 the total reached 3,331,344 and in 1978 some 6,021,293 newly minted Krugerrands were purchased, although this fell to 4,940,755 in 1979. The sales figures for 1980 were 2,844,872 one-ounce, 245,096 ½-ounce, 353,916 ¼-ounce and 856,011 one tenth-ounce Krugerrands. Between 1970 and 1978 the total quantity of gold sold in the form of Krugerrands has been about equal to a full year of South African gold production. *Gold 1980* shows that in 1979 there was a decline in coin production to 245 tonnes compared with 194 tonnes in 1978. The major markets in 1979 were the USA and Germany. German net imports of gold coin have risen particularly steeply in recent years, namely from 28 tonnes in 1973 to 70 tonnes in 1978 and 91 tonnes in 1979. Two-thirds of the imports of gold coins come from South Africa, the Krugerrand being especially popular. Over half of the remainder is accounted for by Canada, imports of whose Maple Leaf are particularly large. Towards the end of 1979 the strong German demand for gold coins was partly due to the fact that as from January 1, 1980, turnover in those gold coins which are formally regarded as legal tender and had been exempt from value added tax until then became subject to that tax, like all other turnover in gold coins and gold bullion.

The procedure followed in the production of Krugerrands is probably typical of that used in the production of other gold coins. In this, electrolytic gold is melted in an induction furnace and cathode copper added in requisite amount afterwards, to produce an Au-Cu alloy containing 91.63 to 91.67 per cent Au. This alloy is then poured into vertically closed moulds to produce billets 12 mm. thick and the billets are then rolled to the requisite thickness, being reduced initially by 1.5 mm per pass through the rolls. When the billet is almost to size, it is passed several times at the same setting to ensure maximum uniformity and is then punched to produce coin blanks which are slightly overweight. These are tumbled in soap solution to remove sharp edges and to clean their surfaces. These blanks are then sorted by weight and those still overweight are ground down to the correct weight by allowing them to roll diagonally across a fast moving endless belt of aluminium oxide paper. The finished blanks are then ready for the minting process.

Gold 1981 stresses that 1980, like 1979, was a record year for gold coin production in the United Kingdom. Gold use in the fabrication of proof sovereigns and gold coins for foreign countries gives a total for 1978 of 59 tonnes. This can be compared with the figure of 0.7 tonnes only ten years earlier. Demand for the Mexico Peso remained high in 1979 and the Government Mint doubled its output from 23 tonnes in 1978 to 46 tonnes in 1979.

Stockpiling and Speculation

In recent years the bulk of gold purchases for private stockpiling has been motivated by expectations that gold will appreciate in terms of money. This motive holds true even for those who have bought gold as a safe investment or as a store of value, since such confidence implies that gold is expected to maintain its value more securely than money or securities. Most of these buyers of gold, seeking merely protection against inflation, do not regard themselves as speculators; but for purposes of analysis it makes no difference whether a purchaser wants to avoid loss from depreciation of money or whether he seeks capital gain from the appreciation of gold. Being more conscious of seeking safety from loss through inflation, the speculator usually has no specific expectations as to the timing and extent of the increase in the money value of gold. While believing that gold is safer than currencies or securities, he does not speculate about whether its appreciation will begin next month or next year and whether it will occur in one fell swoop, or in several smaller jumps.

These types of buyers are influenced chiefly by rumours, news reports, political developments, particularly in the Middle East and South Africa, the price of energy, and by financial analysts' newsletters. They buy gold because they have been told that its price will be doubled, or tripled, either because gold production is going to be reduced and/or gold consumption is increasing at a fabulous rate each year, or because all other commodities have had their prices more than doubled in recent years and the price of gold will necessarily catch up with the rest. Whatever the reason they expect an increase in the price of gold and they do not want to miss the chance of a good profit.

Hoarding of gold is fundamentally different from speculation in gold. Hoarding for *safekeeping* involves the holding of part of one's personal wealth in the form of gold, independently of price expectations other than for at least a reasonably stable price. The motivations are diverse, such as the convenience derived from the concentration of considerable wealth (which makes it easier to hide and transport secretly), more often with the purpose of avoiding taxation or confiscation than for prevention of loss or theft.

Hoarding of gold, silver, other precious metals, and gems for *traditional reasons* reflects a way of life in the Near, Middle and Far East. The predominant factor seems to be the lack of readily available alternatives for storing one's wealth, such as local bank accounts, foreign bank accounts, and local securities markets. Even as new opportunities for investment do become available tradition may slow the transition to these new forms. Traditions such as inheritance laws and the right of a woman to own and hold precious metals and gems (separately from her husband's property) often play a major role. Local customs and social prestige often place emphasis upon visible wealth in the form of ornate to very crude jewellery,

Table 5.6. Gold in Official Coins (net).
(Metric tons)

	1970	1971	1972	1973	1974	1975	1976	1977	1978	1979	1980
South Africa	7.2	17.8	21.7	30.1	99.7	173.6	90.9	88.9	193.6	145.4	107.1
Mexico	—	—	10.0	12.0	66.7	9.9	14.5	0.8	22.7	45.6	23.0
United Kingdom	1.6	0.5	0.5	0.7	14.7	29.3	8.1	22.8	49.1	57.6	59.0
Turkey	12.7	9.0	7.4	3.4	6.0	13.5	14.3	11.0	5.5	—	1.6
Austria	22.4	23.3	20.7	4.4	75.0	9.5	28.4	7.8	5.1	0.6	0.1
Canada	—	—	—	1.0	0.1	0.3	11.1	3.0	4.6	35.0	42.8
Iran	1.0	1.0	0.5	1.0	3.0	3.0	4.0	4.0	3.2	−1.0	−5.0
Switzerland	0.5	0.2	0.1	—	—	0.8	0.4	0.5	1.2	1.0	0.6
United States	—	—	—	—	—	2.7	1.4	1.0	1.0	1.0	0.5
Italy	—	—	—	—	1.0	1.0	0.7	0.5	0.7	0.7	0.7
Chile	0.5	—	—	—	2.0	1.5	3.6	0.6	—	1.7	1.3
Netherlands	—	—	0.1	—	—	1.9	0.2	0.1	—	0.2	0.1
Thailand	—	1.7	—	—	—	0.1	—	0.4	—	—	—
Australia	—	—	—	—	—	—	0.1	0.1	0.4	—	—
Belgium	—	—	—	—	—	—	3.0	—	0.3	—	2.3
Brazil	—	—	0.5	—	—	—	—	—	0.1	—	—
Colombia	—	—	—	1.5	—	—	—	—	—	—	—
France	0.5	0.5	0.5	—	0.4	—	—	—	—	—	—
Israel	—	—	—	—	1.3	—	1.3	0.1	—	—	—
Malta	—	—	0.8	0.2	0.2	—	0.2	—	—	—	—
Peru	—	—	—	—	—	—	—	—	—	1.3	—
Singapore	—	—	—	—	—	2.1	—	—	—	—	—
Yugoslavia	—	—	—	—	—	—	—	—	—	0.6	—
Hungary (see below)	—	—	—	—	17.0	—	—	—	—	—	—
Total	46.4	54.0	62.8	54.3	287.1	251.1	182.2	141.8	287.5	289.7	234.1

Hungary: Coins made from gold purchased from the private sector. Subsequently these coins were sold into the private sector.

Source: Gold 1981 published by Consolidated Gold Fields Limited.

with less prestige being associated with less visible forms of wealth such as bank deposits, stocks and bonds.

It is generally presumed that both types of hoarding are relatively insensitive to either the price of gold or to changes in the price of gold, in contrast to speculation which is very sensitive to actual and expected changes in the price of gold. Hoarding seems to be influenced primarily by factors outside the gold market, so the rate of hoarding or dishoarding is relatively independent of the price in the gold market.

Hoarding of gold for purposes of safekeeping rests on the fact that a large value in terms of money is contained in a small physical volume and therefore can be more easily hidden or guarded from theft and/or confiscation. Still, there must be also an assumption that gold will not drastically depreciate and thus that gold will satisfactorily serve the 'store of value' function of an asset. Hence, expectations about the future price of gold are tacitly presupposed even on the part of safekeeping hoarders. Expectations of a highly unstable and possible lower price for gold would significantly reduce this category of gold-hoarding by encouraging the hoarding of substitutes with more attractive prospects.

In practice the speculative and hoarding motives are impossible to disentangle. These two motives are lumped together by Consolidated Gold Fields Limited under the heading 'bullion holdings'. This total is calculated by estimating the supply of gold to the non-communist private sector and subtracting from it the total for gold fabrication. As can be seen from Table 5.2 there has been considerable variability in this item. The net private bullion purchases of 282 tonnes in 1980 is made up of gold which is physically transferred to the country of residence of its new owners. The remainder stays in the country where it was purchased because the new owner prefers this arrangement. Estimates of the part which is moved and which is to be used for hoarding and investment are made by Consolidated Gold Fields Limited by adding the total flow of gold into each country to any domestic production and subtracting re-exports and fabrications. The result of this calculation will be positive when there is an increase in physical gold for hoarding and investment but it will be negative when there is dishoarding and disinvestment.

During 1979 *Gold 1980* states there was a major increase in the physical gold which was transferred in this manner. Taiwan was the largest recipient with 68 tonnes followed by Hong Kong and Indonesia with 25 tonnes and 23 tonnes respectively. Political uncertainties in Taiwan have been the motive for the increase in hoarding. These uncertainties were caused by the United States' recognition of Peking and the subsequent exclusion of Taiwan from the IMF. Substantial tonnages were also estimated for Turkey with 16 tonnes, Singapore with 12 tonnes and Thailand with 10 tonnes. Dishoarding took place from Kampuchea, Vietnam, Iran and Saudi Arabia.

Central Bank Gold Holdings

On the demand side, a significant new source of buying pressure has recently been added to the traditional regular purchases by jewellers, dentists, industrial users and private investors. Central banks and government connected institutions from developing countries, ranging from the oil rich Middle East to their poorer relations in South America, have emerged as buyers on the market. These purchases from the Third World have been 'small but regular' according to David Potts, the chief gold analyst of Consolidated Gold Fields Limited. Indonesia purchased 1.49 million ounces (46 tonnes) during the first four months of 1980, while Taiwan, Pakistan, the Philippines, Colombia, Libya, Iran, Bolivia and Peru all acquired lesser amounts. Indonesia's total purchases amounted to 1.8 million ounces, one of the largest annual purchases of gold by any central bank.

Why do central banks hold non-interest bearing gold? It used to be the case that gold stocks were legally needed to back a country's monetary policy but no country (apart from Switzerland) now statutorily adheres to this type of policy. Other considerations, however, are important and receive differing degrees of stress depending on the traditions of the central banks concerned.

First, central bankers stress that gold is unique in that the asset of the holding country is not a liability of another country. For dollar assets, on the other hand, there must be a liability in the United States, either money market paper or bank deposits. Hence, the disposition of gold is entirely in the hands of the holding country, while the use of dollars may require the acquiescence of the USA. This is the 'war chest' motive, a motive which leads to the belief that gold holdings in some sense increase a country's sovereignty. This motive acquired increased importance when President Carter, under the provisions of the International Emergency Powers Act (1977), announced on 14th November 1979 the freezing of Iranian financial assets held in US banks. This measure led to a complete re-think of the nature of the Eurodollar market, a market which previously was thought to be external to the USA.

A second motive for central bank gold holding is that the foreign exchange risk to a central bank on its gold reserves is limited to a possible fall in the price of gold or to an appreciation of its own currency *vis-à-vis* gold. On dollar reserves there is the additional risk of its own currency appreciating *vis-à-vis* the dollar as a result of a rise in the dollar price of gold while its own price of gold remains unchanged. When sterling depreciated in 1931, some central banks had large balance sheet losses on their sterling holdings caused by the change in exchange rate relationships. The small group of central banks that hold a large part of their reserves in gold are concerned primarily to avoid such a risk to their balance sheet position. They do not consider it a primary function of the central bank to earn interest on its reserves and they believe that having their reserves

Table 5.7. Official Gold Holdings of Central Banks (excluding China, USSR and associated countries).

	30th September, 1980 Tons	SDR's (millions)	End 1979 Tons	SDR's (millions)	End 1978 Tons	SDR's (millions)
USA	8,227	9,257	8,230	9,261	8,597	9,674
Canada	657	739	690	776	688	775
Austria	657	739	657	739	655	737
Belgium	1,063	1,196	1,064	1,197	1,325	1,491
France	2,546	2,865	2,548	2,867	3,172	3,570
German Federal Republic	2,961	3,331	2,963	3,334	3,690	4,152
Italy	2,074	2,334	2,073	2,335	2,585	2,909
Japan	754	848	754	848	746	839
Netherlands	1,367	1,538	1,368	1,539	1,704	1,917
Portugal	689	775	688	774	688	775
Republic of South Africa	374	421	312	351	305	343
Switzerland	2,590	2,914	2,590	2,914	2,590	2,915
UK	584	657	568	639	710	799
OPEC	1,207	1,358	1,138	1,279	1,128	1,269
Other Asia	607	683	595	669	645	726
Other Europe	1,209	1,360	1,200	1,350	1,073	1,207
Other Middle East	461	519	459	516	470	529
Other Western Hemisphere	654	734	621	699	583	656
Rest of World	320	360	318	358	312	351
Unspecified	113	128	92	104	14	17
Total All Countries	29,110	32,756	28,926	32,549	31,680	35,651
IMF	3,217	3,620	3,323	3,739	3,676	4,137
EMCF	2,664	2,998	2,653	2,985	—	—

Based on the market price of US $666.75 on 30th September, 1980, official holdings, including IMF and EMCF, were valued at some US $750,100 millions.

Note: 1 metric ton = 32,151 ounces.

Source: IMF.

Table 5.8. Gold Holdings in Millions of Fine Troy Ounces.

Country	1975	1976	1977	1978	1979	June 1980	July 1980	August 1980
Argentina	4.00	4.00	4.18	4.28	4.37	4.37	n.a.	n.a.
Australia	7.38	7.36	7.65	7.79	7.93	7.93	7.93	7.93
Austria	20.88	20.88	21.00	21.05	21.11	21.11	21.11	21.11
Belgium	42.17	42.17	42.45	42.59	34.21	34.18	34.18	34.18
Brazil	1.33	1.33	1.52	1.61	1.70	1.70	n.a.	n.a.
Canada	21.95	21.62	22.01	22.13	22.18	21.31	21.30	21.22
Chile	1.30	1.34	1.36	1.39	1.52	1.62	1.67	1.67
Denmark	1.81	1.81	1.93	1.98	1.64	1.63	1.63	1.63
Egypt	2.43	2.43	2.43	2.47	2.47	2.43	n.a.	n.a.
Finland	0.82	0.82	0.91	0.95	0.99	0.99	0.99	0.99
France	100.93	101.02	101.67	101.99	81.92	81.85	81.85	81.85
Greece	3.63	3.65	3.73	3.77	3.81	3.81	3.82	n.a.
Iceland	0.03	0.03	0.04	0.04	0.05	0.05	0.05	0.05
India	6.95	6.95	7.36	8.36	8.56	8.56	n.a.	n.a.
Indonesia	0.06	0.06	0.17	0.22	0.28	1.27	1.40	1.62
Ireland	0.45	0.45	0.47	0.45	0.38	0.36	0.36	0.36
Israel	1.10	1.10	1.16	1.17	1.23	1.23	1.18	n.a.
Italy	82.48	82.48	82.91	83.12	66.71	66.67	66.67	66.67
Japan	21.11	21.11	21.62	23.97	24.23	24.23	24.23	24.23
Malaysia	1.66	1.66	1.74	1.89	2.13	2.29	n.a.	n.a.
Mexico	3.66	1.60	1.76	1.89	1.98	n.a.	n.a.	n.a.
Netherlands	54.33	54.33	54.63	54.78	43.97	43.94	43.94	43.94
New Zealand	0.02	0.02	0.04	0.07	0.05	0.02	0.02	0.02
Nigeria	0.57	0.57	0.63	0.63	0.63	0.63	0.63	n.a.
Norway	0.98	0.98	1.08	1.13	1.18	1.18	1.18	1.18

Table 5.8. (continued).

Country	1975	1976	1977	1978	1979	June 1980	July 1980	August 1980
Pakistan	1.59	1.62	1.62	1.72	1.77	1.82	1.82	1.82
Philippines	1.06	1.06	1.06	1.51	1.70	1.79	1.80	1.82
Portugal	27.72	27.67	24.11	22.13	22.13	22.16	22.16	n.a.
South Africa	17.75	12.67	9.72	9.79	10.03	11.34	11.79	11.83
South Korea	0.11	0.11	0.15	0.28	0.30	0.30	0.31	0.30
Spain	14.27	14.27	14.44	14.52	14.61	14.61	14.61	n.a.
Sweden	5.79	5.79	5.93	6.00	6.07	6.07	6.07	6.07
Switzerland	83.20	83.28	83.28	83.28	83.28	83.28	83.28	83.28
Turkey	3.57	3.57	3.63	3.67	3.77	3.77	3.77	n.a.
United Kingdom	21.03	21.03	22.22	22.83	18.25	18.75	18.76	18.76
United States	274.71	274.68	277.55	276.41	264.60	264.60	264.60	264.60
Venezuela	11.18	11.18	11.25	11.39	11.36	11.46	11.46	11.46
West Germany	117.61	117.61	118.30	118.64	95.25	95.18	95.18	95.18

Source: *International Financial Statistics.*

in gold gives them greater independence of action in monetary crises.

Another consideration in central bank reserve policy is the possibility of a universal rise in the price of gold as occurred in early 1980. A central bank which holds only dollars would not have the benefit of the marked up value of reserves that gold holding countries would have. In a number of countries, especially in Europe, the attachment of governments and central banks to gold is also motivated by the desire to display respectably large gold stocks to people who, having lived through the inflation of the past half century, keep an eye on the state of the nation's monetary reserves as a vital indicator of the soundness of its domestic finances.

The rise in the gold price throughout 1979 has had two important repercussions on the market. It has lowered the readiness of official holders to dispose of their stocks, and has increased demand from countries with low gold reserves, in the Middle East and elsewhere, to build up their holdings. Other monetary consequences of the rise in the gold price and the impact this has on central banks were discussed in Chapter 4.

As can be seen from Table 5.7, at the end of September 1980, the USA had around 8227 tonnes of gold, considerably more than the IMF (3217 tonnes), the European Monetary System members (2664 tonnes), West Germany (2961 tonnes), Switzerland (2590 tonnes), France (2546 tonnes), Italy 2074 tonnes), The Netherlands (1367 tonnes) and OPEC (1207 tonnes). The USA holding is approximately 28 per cent of the total gold held by the non-communist countries and it continues to value it at $42.22 per ounce. Table 5.8 gives gold holdings expressed in fine Troy ounces.

The Supply of Gold

The world's supply of gold comes from five sources. These are current production, reprocessed scrap, a reduction of processor's inventories, reductions of other private stocks and reductions of monetary stocks. Current production in turn comes from old mines, newer mines, and improved mining techniques. Gold is unique among all commodities with respect to the vast quantity that is stockpiled. If all production of new gold and reprocessing of old gold were to cease completely it would take over one hundred years for the current stocks to be consumed at the current rate.

David Potts estimates that some 100,000 tonnes of gold (3,215 million ounces) have been mined since anyone first thought the metal worth digging up. In addition, official gold holdings are equivalent to some 25 years production at current rates. The South African Gold Coin Exchange estimates gold production at 26 tonnes in the sixteenth century, 45 tonnes in the seventeenth century, 90 tonnes in the eighteenth century, 4864 tonnes in the nineteenth century and (up until 1977) 77,393 tonnes in the twentieth century.

Table 5.9. World Gold Production.
(Thousand Fine Troy Ounces)

Year	Ounces	Year	Ounces	Year	Ounces
1801–10*	608	1902	14,494	1940	42,176
1810–20*	391	1903	15,934	1941	39,030
1820–30*	486	1904	16,920	1942	35,325
1830–40*	652	1905	18,488	1943	27,989
1840–50*	1,760	1906	19,534	1944	25,346
1850–60*	6,445	1907	20,040	1945	24,483
1860–70*	6,107	1908	21,484	1946	24,946
1871	6,381	1909	22,094	1947	25,347
1872	5,798	1910	22,147	1948	26,559
1873	5,504	1911	22,475	1949	27,580
1874	5,360	1912	22,637	1950	28,257
1875	5,341	1913	22,352	1951	27,373
1876	5,430	1914	21,218	1952	27,915
1877	6,001	1915	22,649	1953	27,775
1878	5,987	1916	22,047	1954	29,133
1879	5,416	1917	20,216	1955	30,441
1880	5,349	1918	18,523	1956	31,454
1881	5,064	1919	17,543	1957	32,644
1882	4,886	1920	16,304	1958	33,676
1883	4,746	1921	15,987	1959	36,152
1884	5,015	1922	15,471	1960	37,809
1885	5,102	1923	17,781	1961	39,294
1886	4,945	1924	19,031	1962	41,600
1887	5,256	1925	19,013	1963	43,147
1888	5,509	1926	19,343	1964	44,841
1889	6,048	1927	19,388	1965	46,225
1890	5,814	1928	19,433	1966	46,580
1891	6,400	1929	19,589	1967	45,737
1892	7,060	1930	20,873	1968	46,165
1893	7,544	1931	22,341	1969	46,612
1894	8,657	1932	24,255	1970	47,522
1895	9,518	1933	25,511	1971	46,495
1896	9,717	1934	27,028	1972	44,843
1897	11,397	1935	29,460	1973	42,997
1898	13,921	1936	33,101	1974	40,124
1899	15,073	1937	35,263	1975	38,476
1900	12,421	1938	37,598	1976	39,642
1901	12,692	1939	39,635	1977	39,385

*Annual Average for the Ten-Year Period.
Sources: 1801–1870   American Metal Market, *Metal Statistics 1974*, attributed to Dr. Adolph Soetbeer.
       1871–1977   US Bureau of Mines.

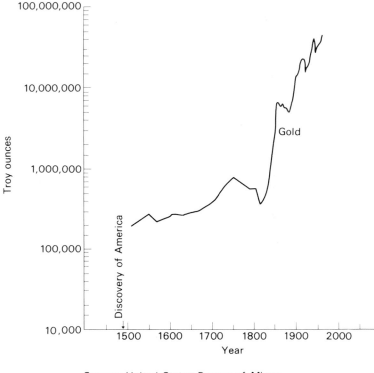

Source: United States Bureau of Mines
Director of the U.S. Mint

Figure 5.1. Trend in world production of gold (annual output).

Production of new gold since the discovery of America has come in a
series of great surges, as discoveries were made and new mines brought
into production. These were discussed in detail in Chapter 2. The first
came in the sixteenth century when the Spaniards seized the treasures
gathered by the Aztecs and Incas in Mexico and Peru. An even greater
surge in gold supply quickly followed the discovery of the 'placers' in
California and in South-Eastern Australia in 1848 and 1851 respectively.
As can be seen from Figure 5.1 the curve of production continued to rise,
with minor fluctuations, until the outbreak of war in 1914. The California
production was supplemented by gold from the many new districts in the
North American Cordillera and the pre-Cambrian shield of Canada, among
which the Comstock, the Black Hills, the Yukon and Alaska, found in the
nineteenth century, and Cripple Creek, Tonopah, and Goldfield, in the

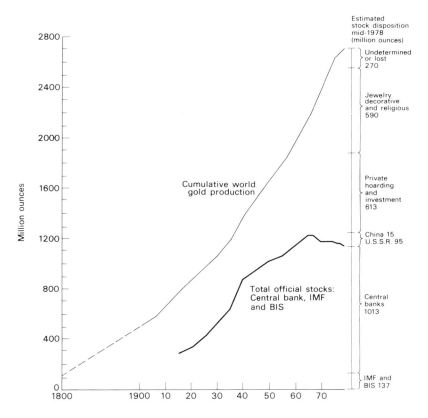

Data: U.S. Bureau of Mines, International Monetary Fund, Charter Consolidated, and J. Aron Precious Metals Research Department

Chart prepared by J. Aron Precious Metals Research Department

Figure 5.2. Cumulative world gold production and its distribution.

present century, as well as Porcupine, Kirkland Lake, and many other Canadian districts, deserve mention. Gold also was supplied by many new mines in Western Australia, the Philippines, the Gold Coast, and elsewhere in this period. The outstanding event, however, was the discovery of gold in the Transvaal in 1886 and the development of a succession of mines in this immense region that before the turn of the century made it by far the major gold field of the world, a position it still holds today.

Gold has been discovered in all five Continents of the world. In the

Table 5.10. The increase in the gold holdings of the
non-communist private sector (metric tons).

	Non-communist world mine production	Net trade with com-munist sector	Net official sales (+)	Net official purchases (−)	Total
1948	702	—	—	369	333
1949	733	—	—	396	337
1950	755	—	—	288	467
1951	733	—	—	235	498
1952	755	—	—	205	550
1953	755	67	—	404	418
1954	795	67	—	595	267
1955	835	67	—	591	311
1956	871	133	—	435	569
1957	906	231	—	614	523
1958	933	196	—	605	524
1959	1000	266	—	671	595
1960	1049	177	—	262	964
1961	1080	266	—	538	808
1962	1155	178	—	329	1004
1963	1204	489	—	729	964
1964	1249	400	—	631	1018
1965	1280	355	—	196	1439
1966	1285	−67	40	—	1258
1967	1250	−5	1404	—	2649
1968	1245	−29	620	—	1836
1969	1252	−15	—	90	1147
1970	1273	−3	—	236	1034
1971	1236	54	6	—	1386
1972	1182	213	—	151	1244
1973	1117	275	66	—	1398
1974	1006	220	20	—	1246
1975	954	149	9	—	1112
1976	970	412	58	—	1440
1977	972	401	269	—	1642
1978	979	410	362	—	1751
1979	961	199	544	—	1704
1980	943	90	—	230	803

Definition of official sales has been extended from 1974 to include activities of government controlled investment and monetary agencies in addition to central bank operations. This category also includes IMF disposals.
Source: *Gold 1981* published by Consolidated Gold Fields Limited.

Americas its distribution lies mainly down the west coast, from Alaska in the north down through the great extruded mass of the Rockies into Mexico and the northern half of South America: a secondary gold producing area can be defined southwards from Quebec to the South-Eastern area of the USA. The principal European deposits have extended from the British Isles south-eastward through France and along the line of the Alpine and Balkan mountain ranges, together with North-Western, Central and Southern Iberia. In Asia an immensely extended belt of gold deposits has lain across the centre of Asiatic Russia: in addition Anatolia, Armenia, Arabia, Southern and Eastern India and the Hindu-Kush have produced gold, together with Korea, the Chinese littoral and Japan. The pattern of these latter deposits spreads downwards through the Eastern archipelago into Australia and New Zealand. Finally, the African Continent which has produced gold in its central and western zones as well as the regions of Upper Egypt, Nubia and Ethiopia and the Southern plateau.

The figures for world gold production and cumulative world gold production are given in Table 5.9 and Figure 5.2. Consolidated Gold Fields Limited break down the sales of new gold to the non-communist private sector into three categories. First, the non-communist private world mine production, second sales (net of purchases) from the communist bloc and thirdly sales (net) and purchases (net) of official international financial organisations, e.g. the IMF. The amounts of gold supplied by each component for the years 1948 to 1980 are given in Table 5.10.

Non-communist world gold mined can be seen from Table 5.10 to have been at its highest in the 1960s and early 1970s but has slowly declined throughout the later 1970s. Net trade with the communist bloc follows no consistent pattern although it is widely felt that Soviet gold sales are purely for the purpose of obtaining foreign exchange, so that physical deliveries of gold from that source rise and fall in relation to the bullion price. Net official sales first started in 1966 and built up particularly for the years 1977, 1978 and 1979. Net official purchases which ceased after 1972 restarted in 1980.

Comparisons between the data for 1980 and 1979 show that there was little change in non-communist world mine production but that for the second year running there was a significant fall in the net sales from the communist bloc. The total supplied from the three categories was 803 tonnes in 1980 compared with 1704 tonnes in 1979. This fall was largely due to the cessation of official sales.

Non-Communist World Production

Geographically the world's gold supply is dominated by production from two countries, South Africa and the USSR. Table 5.11 gives details of gold production in the non-communist world.

South Africa as a Gold Producer

Gold occurs in South Africa in native form deposited on quartz pebbles cemented together by more quartz to form hard conglomerates known as reefs. The reefs occur in layers and dip at varying angles and are of thickness varying between a few inches and several feet. All the mining is underground (as opposed to open cut) and at the present time the reef is extracted by blasting horizontal sections of rock about 4 feet high which include the reef. The material thus brought down is then hauled to the surface and any that is obviously not reef is removed. The remainder is sent to the reduction plant for the gold to be extracted.

As can be seen from Tables 5.11 and 5.12 South Africa has consistently dominated the gold supply which is derived from the non-communist world. As a percentage of total non-communist gold production, three-quarters of it has been supplied by South Africa with the proportion falling slightly from 77 per cent in 1969 to 73 per cent in 1979 (see Table 5.12). The 1979 South African gold production at 675.0 tonnes was slightly down on the 703.3 tonnes produced in 1979. This 1980 output represents some 50 per cent of total world production. The gold production in 1980 of some 675 tonnes is the lowest output for some 20 years.

South Africa has seven gold fields with 37 major gold mines forming a 'Golden Arc' stretching 500 kilometres across its centre. These are illustrated in Figure 5.3. The gold mines employ about 400,000 men, roughly two out of every three mining men in South Africa. Up to 260,000 men are underground on gold mines during the main shift every day, blasting about eight million tons of rock a month. Of the two major gold mines in South Africa, namely Vaal Reefs and West Driefontein, each produce more gold in a year than Canada. The value of South African gold output in 1979 was some R6 billion and rose to R10 billion in 1980. In recent years the centre of gravity of South African production has shifted away from the West, Central and East Rand areas, and the bulk of current production is now from the Orange Free State, Klerksdorp and Far West Rand gold mining areas.

With 46 per cent of its exports being derived from gold sales in 1979, and just over 50 per cent in 1980, South Africa is not in the same position as the Russians, far less dependent on gold exports, with respect to their ability to manipulate the gold market. The 1979 South African gold production was obtained from 34 gold mines, two mines which produce uranium as their main product and from the treatment of mine dumps. The largest mine produced 67 tonnes of gold, employed 41,000 workers and extracted 8 million tonnes of ore. It thus required 119,400 tonnes of ore to yield one tonne of gold.

South African gold is being mined from lower grades of ore. In 1979 the average grade was 8.2 grams per tonne compared with 9.4 grams per ton in 1978. Between 1972 when the grade was 12.7 grams per tonne and mid-1980 the grade of the ore had fallen by approximately 35 per cent.

The fact that the gold price has risen faster than working costs has enabled the mining of lower grade ores to take place. A second reason for the substitution of lower grade ores for higher grade ores is that some of the major producers in the Orange Free State have now been operating for over twenty years and have exhausted their higher grade ores and the resulting fall in gold production can only be partially offset by increasing the mining and milling capacity.

Great concern has been expressed in the South African gold mining community about rising costs. Table 5.13 indicates the average milling breakeven point for the whole South African mining industry, in terms of grams of gold which have to be produced from a tonne of rock milled in order to pay for the costs of mining and milling that tonne of rock. (The higher the gold price the lower the grams per tonne, g/t, breakeven point and vice versa.)

Table 5.13 highlights a very interesting fact that, although the breakeven point has fluctuated quite widely, it has oscillated essentially around approximately 5 g/t: in 1933 the breakeven point was 5.3 g/t. In 1948, before the devaluation of sterling, it was 5.2 g/t. By 1953, the effect of the benefits of the devaluation of sterling had been whittled away by cost increases and the breakeven point had returned to 5.1 g/t. Subsequently, the breakeven point rose steadily until the establishment of the two-tiered gold market in 1968. By 1973, the breakeven point had dropped from 8.5 g/t in 1968 to 5.0 g/t. Thereafter it dropped to 3.8 g/t in 1974, when the gold price surged ahead prior to the legislation of gold ownership in the USA. The breakeven point climbed steadily to 5.9 g/t in 1977, a particularly difficult year for the industry. In 1978 the breakeven point returned to 5.0 g/t which, incidentally, is the average breakeven point for the last 6 years in the table. Looking at these figures, it is clear that the price of gold, in South African currency terms, became progressively more undervalued in the 15 years up to 1968. In the 5 years to 1973 there was a major structural readjustment in the gold mining industry. With the gold price rising and costs under proper control, the breakeven point may fall.

The South African government has an obvious interest in extending the lives of the mines to maintain taxation revenues, employment levels and gold's contribution to the South African balance of payments, and the mines have to conduct their operations to meet these policy objectives. Consequently, the mines have to extract lower grade ore after a significant and long-term rise in the gold price. This objective is achieved by a term in the mining lease which requires a mine to calculate the average value of its payable reserves. The mine plans are then designed to ensure that the grade of ore extracted is close to the average. Payable reserves are defined as those which have a higher grade than the pay limit which in turn crucially depends upon assumptions about future revenues and costs. Given the high cost of opening and closing a mine, the mining industry tends to be conservative as to its future gold price assumptions.

Table 5.11. Gold Production in the Non-communist World.
(Metric tons)

	1970	1971	1972	1973	1974	1975	1976	1977	1978	1979	1980
South Africa	1000.4	976.3	909.6	855.2	758.6	713.4	713.4	699.9	706.4	703.3	675.0
Canada	74.9	68.7	64.7	60.0	52.2	51.4	52.4	54.0	54.0	51.1	49.3
USA	54.2	46.4	45.1	36.2	35.1	32.4	32.2	32.0	30.2	30.2	27.6
Other Africa											
Zimbabwe	15.0	15.0	15.6	15.6	18.6	18.6	17.1	20.0	17.0	12.0	11.4
Ghana	21.9	21.7	22.5	25.0	19.1	16.3	16.6	16.9	14.2	11.5	12.8
Zaire	5.5	5.4	2.5	2.5	4.4	3.6	4.0	3.0	1.0	2.3	3.0
Other	2.0	2.5	1.7	1.7	1.5	1.5	1.5	1.5	2.0	2.5	2.5
Total Other Africa	44.4	44.6	42.3	44.8	43.6	40.0	39.2	41.4	34.2	28.3	29.7
Latin America											
Brazil	9.0	9.0	9.5	11.0	13.8	12.5	13.6	15.9	22.0	25.0	35.0
Dominican Republic	—	—	—	—	—	3.0	12.7	10.7	10.8	11.0	11.5
Colombia	6.8	5.9	6.3	6.7	8.2	10.8	10.3	9.2	9.0	10.0	16.6
Mexico	6.2	4.7	4.6	4.2	3.9	4.7	5.4	6.7	6.2	5.5	5.9
Peru	3.2	3.0	2.6	2.6	2.7	2.9	3.0	3.4	3.9	4.7	5.0
Nicaragua	3.6	3.3	2.8	2.8	2.4	1.9	2.0	2.0	2.3	1.9	1.5
Other	6.6	8.2	9.0	7.9	5.9	6.0	8.0	8.0	8.5	8.5	10.0
Total Latin America	35.4	34.1	34.8	35.2	36.9	41.8	55.0	55.9	62.7	66.6	85.5

Table 5.11. (continuation).

	1970	1971	1972	1973	1974	1975	1976	1977	1978	1979	1980
Asia											
Philippines	18.7	19.7	18.9	18.1	17.3	16.1	16.3	19.4	20.2	19.1	22.0
Japan	8.2	8.2	7.8	6.2	4.5	4.7	4.5	4.8	4.7	4.2	3.4
India	3.2	3.7	3.3	3.3	3.2	3.0	3.3	2.9	2.8	2.7	3.0
Other	2.8	2.1	2.7	2.7	2.7	2.7	3.0	3.0	3.0	3.0	3.0
Total Asia	32.9	33.7	32.7	30.3	27.7	26.5	27.1	30.1	30.7	29.0	31.4
Europe	7.4	7.6	13.2	14.3	11.6	11.0	11.4	13.2	12.5	10.0	9.2
Oceania											
Papua/New Guinea	0.7	0.7	12.7	20.3	20.5	17.9	20.5	22.3	23.4	19.7	14.0
Australia	19.5	20.9	23.5	17.2	16.2	16.3	15.4	19.2	20.1	18.6	17.3
Other	3.6	3.1	3.2	3.2	3.2	3.2	3.0	4.0	4.7	4.5	4.0
Total Oceania	23.8	24.7	39.4	40.7	39.9	37.4	38.9	45.5	48.2	42.8	35.3
Non-communist world total	1273.4	1236.1	1181.8	1116.7	1005.6	953.9	969.6	972.0	978.9	961.3	943.0

Source: *Gold 1981* published by Consolidated Gold Fields Limited.

Table 5.12. A Country Break-down of Non-communist Gold Production expressed in Percentage Form.

	1969	1970	1971	1972	1973	1974	1975	1976	1977	1978	1979
South Africa	77	78	79	77	76	75	75	74	72	72	73
Canada	6	6	6	5	5	5	5	5	5½	5½	5
USA	4	4	3½	4	3	3	3	3	3	3½	3
Total Other Africa	4	4	3½	4	4	4	4	4	4	3½	3
Latin America	3	3	2½	3	3	4½	4½	6	6	6	7
Asia	3	3	2½	3	3	3½	3½	3	3½	3	3
Europe	1	½	1	1	1	1	1	1	1	1½	1
Oceania	2	2½	2	3	4	4	4	4	5	5	4

Source: Derived from *Gold 1980* Consolidated Gold Fields Limited.

As with the Soviet Union, there has recently been speculation that South Africa has been selling gold for oil to the Middle East countries. Commercially, such deals would make sense. South Africa is officially boycotted by most oil producers. Since the cut-off of Iranian supplies, particularly true since the Iraq/Iranian war, it has at times found it difficult to secure reliable sources of oil.

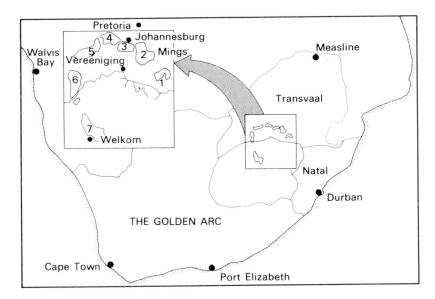

South Africa The Seven Gold Mining Areas

1. Evander (4 mines)   5. Far West Rand (9 mines)
2. East Rand (6 mines)   6. Klerksdorp (4 mines)
3. Central Rand (5 mines)   7. Orange Free State (10 mines)
4. West Rand (2 mines)

Source: South African Chamber of Mines

Figure 5.3. The seven gold mining areas in South Africa.

The South African mining industry is organised around a system known as the 'Group System'. This system developed around finance corporations, each of which administers a group of mining and other companies. The finance corporation and the companies it controls are called a 'group' which takes the name of the controlling or parent corporation.

The six major groups associated together in the Chamber of Mines of South Africa are: Anglo American Corporation of South Africa, Anglo-Transvaal Consolidated Investment Company (Anglovaal), General Mining and Union Corporation (Gencor); Gold Fields of South Africa (GFSA), Johannesburg Consolidated Investment Company (JCI) and Rand Mines. The mining houses are considerably interlinked. An example of this is the owning by De Beers and its associate company Anglo American of 49.9 per cent of JCI and their strategic holding of 25 per cent in Consolidated Gold Fields Limited giving them a powerful presence in GFSA. Consolidated Gold Fields Limited itself owns 46 per cent of GFSA. GFSA

Table 5.13. Gold Mining Breakeven Points

Year	Mill Breakeven Point (g/t)
1933	5.3
1938	4.6
1943	4.5
1948	5.2
1953	5.1
1958	6.5
1963	7.3
1968	8.5
1973	5.0
1974	3.8
1975	4.7
1976	5.8
1977	5.9
1978	5.0
Low Points	
1940	4.2
1950	4.1
1974	3.8

Source: Consolidated Gold Fields Limited.

produce 17 per cent of the free world's gold supply. As Table 5.14 shows Anglo American is by far the biggest mining house in terms of profit and production. Rand Mines is part of the Barlow Rand conglomerate in which Anglo-American has a 7.2 per cent stake.

Within the group each company is a separate entity, with its own body of shareholders and its own board of directors, but the finance corporation provides the companies collectively with technical, secretarial, accounting, buying and other services. Each major mining group has its own consulting, engineering, research, geological and other similar technical departments, and is able to provide sums of capital far beyond the resources of

Table 5.14. Mining Houses' Share of
Gold Production (1979 figures).

	Output tonnes	% of Total
Anglo American	262.00	38.2
Gold Fields of SA	162.70	23.7
General Mining and		
Union Corporation	106.6	15.6
Rand	69.40	10.1
JCI	46.6	6.8
Anglovaal	37.9	5.5
Independent	1.2	0.1
	686.4	100.0

Source: Adapted from *The Economist*,
March 29th, 1980.

the individual companies. The mining groups work closely through the Chamber of Mines of South Africa. The Chamber of Mines is a central cooperative organisation which looks after those interests of its members which can be best handled on a cooperative basis. The members of this 'co-op' control the Chamber of Mines. Associate or subsidiary companies of the Chamber of Mines include the International Gold Corporation (usually known as Intergold) which promotes the sale of gold in the world's leading markets by stimulating consumer demand for gold jewellery. Intergold do this by disseminating information about the qualities of gold and by identifying new uses for gold in science and technology. Intergold also has responsibility for the marketing and promotion of the world's most sold gold coin, the Krugerrand as well as the new mini-Krugerrands. In addition, it controls the Rand Refinery which refines all South Africa's gold output, and which is the largest gold refinery in the world.

The United States and Canada as Gold Producers

The second largest non-communist gold producer is Canada which, as can be seen from Table 5.12, has consistently produced 6 per cent of free world output. Canadian gold production for 1980 has been estimated, by Consolidated Gold Fields Limited, at 49.3 tonnes compared with 51.1 tonnes in 1979. At the end of 1979 there were 21 operational gold mines in Canada where the scale of operation of each mine is very small compared with South African mines. The principal gold mining provinces are Ontario (39 per cent of production), Quebec (29 per cent), British Columbia (16 per cent) and North West Territories (11 per cent).

As can be seen from Figure 5.4, US gold output rose dramatically in the 1850s but since then has fluctuated substantially. Chapter 2 gives details of the Californian gold rush. The US insistence that gold should be demonetised and its insistence that the price should be fixed at $35 per ounce severely penalised the US gold industry. At the beginning of the 1970s only four of the nation's 25 gold producing mines were searching for the metal in a major way. The others were simply producing it as a by-product of other metals. Kennecott Copper, with its major operation in a mountain range near Salt Lake City, produces over 20 per cent of the US gold output as a by-product of copper.

Gold production in the USA has consistently accounted for 3 to 4 per cent of non-communist gold production (see Table 5.12). Production declined in 1980 to 27.6 tonnes from 30.2 tonnes in 1979. Gold was mined in twelve states in 1979 of which South Dakota with its Homestake mine was the leading producer at 7.8 tonnes (251,000 ounces). Second position was held by Utah where gold is produced as a by-product of the Bingham copper mine to a total of 7.4 tonnes. Nevada, the home of the Carlin mine, and Arizona contributed 6.5 and 3.1 tonnes respectively and the four states which have been mentioned account for 88 per cent of US production.

Homestake, which has long been dependent on its original Homestake Mine near Bobtail Gulch in the Black Hills of South Dakota, reputedly discovered by a detachment of General Custer's army in 1874, announced in August 1980 a major new gold discovery. Homestake's new deposit is in California in an area one would have expected to have been well scoured by the pioneers of the 1849 gold rush. Annual output is estimated at a minimum of 100,000 ounces of fine gold over at least the next ten years. The deposit, about 70 miles northwest of Sacramento, in the Napa County, has been graded at about 5 grams of gold per tonne.

Gold 1980 reported a sharp rise in Brazilian gold output in 1979 to an estimated 26.1 tonnes from a revised estimate of 22 tonnes in 1978. Gold 1981 reported a further rise to 35 tonnes in 1980. The increase arose from boosted alluvial gold production in the Amazon region and puts Brazil in fourth place in the non-Communist world production league. The figure of

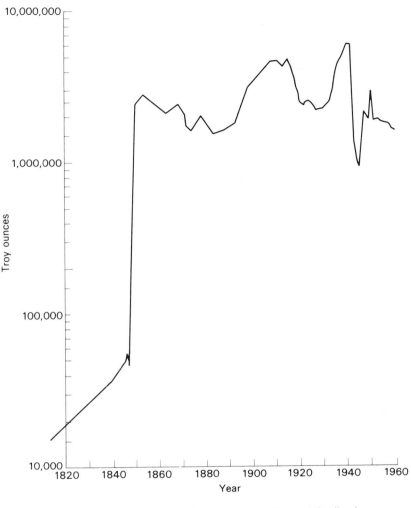

Source: E.M. Wise: *Gold: Recovery, Properties and Applications*,
          Princetown, N.J.: Van Nostrand, 1964

Figure 5.4. US annual production of gold (1820–1960).

35 tonnes in 1980 could rise sharply if the Brazilian government can crack down on smuggling. There are 30,000 gold prospectors, or '*garimpeiros*', in the Amazon basin, the widest gold mine in the world. Of the other

countries, production of Papua New Guinea fell to 14.0 tonnes in 1980 from 19.7 tonnes in 1979 because of a reduction in ore grades at the Rio Tinto-Zinc group's Bougainville mine. Australian gold production fell to 17.3 tonnes in 1980 partly as a result of labour disputes.

The USSR and People's Republic of China as Gold Producers

The subject of gold production in the USSR is a closely guarded State secret with no production figures ever being issued. Westerners have never been allowed to visit Soviet gold producing areas or to speak to those involved in the metal's extraction. Nevertheless, information is obtained via the fact that Russian gold sales normally take place through Zurich and London and by translating and collating the press reports and technical papers which are available in the West.

The main gold producing areas are in Siberia along the banks of the Lena river. The deposits are generally strip-mined using huge machines. But the Russians do allow a system reminiscent of California during the gold rush – individual panning and mining by licensed operators. These miners are generally experienced people who have worked for the State gold mining concern and have acquired enough money to buy bulldozers themselves. Working usually in groups of about 30, they are encouraged to exploit the small deposits that are not economic if mined with the vast strip-mining machinery. The miners have to turn their gold in to the State and are paid according to its value. They work in conditions of isolation and in a harsh, remote landscape, and are usually only able to operate during the summer months.

The post-Second World War emphasis on placer production from the streams and river gravels in the eastern part of the country has swung recently towards important new lode deposits discovered in the South West. By-product gold output is an important source of gold in the USSR, historically accounting for about 20 per cent of total output. The most important single mine in the USSR and the largest in the world, with annual output of around 80 tonnes, is Muruntau, which lies in Uzbekistan. This combine produces something over 20 per cent of total Soviet output.

The People's Republic of China, in contrast to the USSR, is not secret about its gold output. In 1979 a team from Consolidated Gold Fields Limited visited the principal gold mining provinces in the People's Republic and total production is estimated at between 30 and 60 tons a year. The four main gold provinces of China are Shandong, Heilongjuang, Hubei and Guangxi, but new projects are under way in Sichuan, Shaanxi and Yunnan. Indeed, the Chinese claim to have found gold in all the 27 provinces or autonomous regions.

Secondary Sources of Gold

The statistics of sales, purchases and use of gold are usually silent on the question of scrap. However, old scrap is an important above-ground source of gold. Reclamation of gold is difficult as it normally exists in minute quantities and because it is often alloyed with another metal or plated on to a less desirable material. However, because of its high value and the increased cost of primary mining, it is becoming more worthwhile to recycle. For instance, ten years ago when computer parts containing gold were redundant, they were thrown out. Today, this material is being recycled by specialist refiners but not all material can be, or is worth the cost, of reclaiming.

Recycling of gold scrap is carried out in the countries that are the largest consumers, and therefore generators, of scrap, such as North America, Japan and Europe. To understand the source of this scrap, one must look at the major industrial uses which, as mentioned earlier, are in jewellery, electronics, dentistry and coins. Each source has its own scrap forms:

Jewellery: sweepings, grindings, filed scrap, flakes, peelings, punch-outs, solutions, sludges, wire and wipe rags.

Electronics: contacts, diodes, foil, paste, plated parts, wire, solutions, powders, printed circuit boards, punch-outs, components, resins-plating, transistors, wire and wipe rags.

Dentistry: alloys, scrap, sweeps and grinds.

Coins: punch-outs, demonetised coins, etc.

Many of the new high technological uses of gold are for purposes of space exploration and defence and due to the security of these items they may not enter the market for recycling.

Trade with the Communist Bloc

The Communist bloc made substantial sales over the period 1953 to 1965 in order to finance its trade deficit with the West, but withdrew from the market until 1971. Since then, poor harvests and the need to import expensive Western technologies, allied with favourable gold prices, have encouraged the resumption of Soviet sales. Although there is a great deal of secrecy about USSR gold production, for 1979 Consolidated Gold Fields Limited estimate that the net sales of the Communist bloc (principally from the Soviet Union) were 199 tonnes. The 1980 figure of 90 tonnes is 54 per cent below the 199 tonnes produced in 1979. Of the various explanations for this, the most favoured one put forward by western analysts is that the Russians have an annual foreign exchange target and that this was easily achieved in 1980 by a smaller level of gold sales because of the higher prices which Russia was able to enjoy both for gold and oil. An alternative explanation about which there has recently been much speculation, but no proof, is that the USSR has diversified their sales channels and are arranging bilateral deals with oil producing countries. It is certainly the case that the Zurich market, the traditional western outlet for Soviet gold, has no recorded gold sales for the first four months of 1980 although this situation altered in later months.

In addition, there is always the possibility put forward in *Gold 1981* that the Soviet stocks of gold have reached a minimum acceptable level because part of the high level of sales, of over 400 tonnes per year in the three year period from 1976 to 1978, may have come from stocks. Consolidated Gold Fields Limited have estimated annual Soviet gold output at 280 to 350 tonnes per year. If gold consumption of, say, 50 tonnes is subtracted from this, there would be a large gap between official gold sales to the non-communist world and communist production, which must have come from existing stocks.

Official Gold Sales

Table 5.10 indicates the variability that has taken place in net official gold sales. These started in 1966 and have continued through to 1979 with no sales occurring in 1969, 1970, 1972 and 1980. Official gold sales have been undertaken by the US Treasury and the International Monetary Fund.

US Treasury Gold Sales

In an attempt to downgrade the role of gold, the US Treasury announced on 20th April, 1978 that it would hold at least six-monthly auctions at each of which 300,000 ounces (9.33 tonnes) of gold would be offered, the date for the first auction being fixed at 23rd May, 1978. During 1978 some 126 tonnes were sold. The US Treasury held auctions in 1979 of varying amounts of gold in each month except December selling in total 365 tonnes (11.75 million ounces). At the auctions between January and April 1979 the amount on offer was 46.66 tonnes (1.5 million ounces). Of this two-thirds was in the form of gold 995 fine, and the remainder of refinable quality 900 fine. On 18th April it was announced that monthly offerings would be reduced to 23.33 tonnes (750,000 ounces) until further notice and that all gold offered would be of the lower quality. On 16th October 1979, it was announced that in future, in order to frustrate gold speculators, auctions would no longer be held on a regular monthly basis, but that sales might be announced at short notice and for varying quantities. The first and only sale of this type (up to August 1981) was held on 1st November 1979, when 38.88 tonnes (1.25 million ounces) were offered. Table 5.15 gives details of US Treasury auctions held in 1979.

The competitive bidding system was used at all sales during 1979, but there was a change in procedure starting with the auction held in May. Prior to this, all bids had been publicly announced in the order that bid envelopes were opened. This, it was felt, had a short-term disruptive effect on US gold futures markets, which were trading during the opening of bids. The new procedure called for bids to be opened in private, and for

Table 5.15. US Treasury gold sales held in 1979.

Date	Amount offered and sold (in tons)		Average price		Total amount bid for (in tons) (both qualities)	Number of successful bidders (both qualities)
	995 fine	900 fine	995 fine	900 fine		
16.1.79	31.10	15.55	219.71	218.22	211.50	32
22.2.79	31.10	15.55	252.38	251.42	102.64	16
20.3.79	31.10	15.55	241.30	240.09	90.20	28
17.4.79	31.10	15.55	230.96	230.17	102.64	29
15.5.79	—	23.33	—	254.92	74.65	6
19.6.79	—	23.33	—	279.02	62.21	16
17.7.79	—	23.33	—	296.44	65.32	10
21.8.79	—	23.33	—	301.08	71.54	3
18.9.79	—	23.33	—	377.78	80.87	4
16.10.79	—	23.33	—	391.98	37.32	8
1.11.79	—	38.88	—	372.30	47.59	11

Source: Samuel Montagu Annual Bullion Review 1979

the result of the auction to be announced when all bids had been considered, and successful bidders notified.

The most pressing reason for the US decision to make regular sales of gold was the need to improve the country's trade deficit. Many, indeed most, of the buyers at its auctions have been foreign banks and dealers, most notably Dresdner Bank of Frankfurt, which spectacularly scooped 96 per cent, or 720,000 ounces of the gold sold in the Treasury's August 1979 auction. The volume of Treasury sales has probably turned the USA into a temporary net exporter of gold and has certainly reduced bullion imports into the USA which were down from 7.9 million ounces of gold in 1977 to 4.4 million ounces in 1978.

A second stated aim of the Carter administration, like its predecessors, was to downgrade the role of gold in the world monetary system by showing that the USA set no special store by its gold hoard and was willing to sell it off like any other commodity.

The core of the US objection to gold, already discussed in Chapter 4, is that its supply is too unstable and volatile to provide a sure basis for expanding world liquidity. Production and mining of gold is fairly reliable, but with large fluctuations in commercial and industrial use of gold, the amounts left for monetary use, and speculators, it is argued, are difficult to predict.

IMF Gold Sales

Despite the fact that in July 1974 the value of the SDR was changed from an imputed gold value to the combined market value of a 'basket' of sixteen major currencies, weighted by the relative importance of the issuing country in international transactions, the IMF altered its hostility to gold by the ratification on the 1st April, 1978 of the Amendments to the IMF Articles relating to gold. The change was described in the 1978 Annual Report of the Fund:

'Members may no longer define for purposes of the Fund the exchange value of their currency in terms of gold; the official price of gold (SDR 35 per fine ounce) has been abolished, and members are free to trade in and account for gold at any price consistent with their domestic legislation. Members are no longer required to pay gold to the Fund in connection with any transaction or operation, and the Fund may accept gold from a member only by a decision that requires a majority of 85% of the total voting power . . .'

In 1976 the IMF announced its decision to sell one sixth of its gold to the market over a four year period. (This represents some 8 per cent of total gold supplies.) In August 1976 the IMF Interim Committee announced that, as part of a package to reform the monetary system it had agreed to abolish the official price of gold, as mentioned above, and to authorise the IMF to reduce its gold holdings by one sixth (25 million ounces) by selling at market related prices and to make a restricted distribution of a further 25 million ounces (1,555 tonnes in total). The first six IMF auctions were for 780,000 ounces, but it was decided that subsequent auctions would be for 525,000 ounces. Auction numbers three, four, seven, eight and nine were based on a Dutch auction or common price basis, whilst later auctions were based on a multiple bid price arrangement.

The first auction was held on 3rd May, 1978. During 1978 the IMF held twelve auctions at which 184 tonnes were sold to the public. (This is slightly less than the 188 tonnes sold by the IMF in 1977.) In addition 43 tonnes were sold to the Central Banks of developing countries on a non-competitive basis at the auctions conducted between June and December. The IMF again held monthly auctions in 1979 (see Table 5.16). A total of 169.8 tonnes (5.48 million ounces) was sold to the public. The competitive bidding system was used throughout. At the first five auctions the amount sold was 14.62 tonnes (470,000 ounces). During this period also, a further 2.97 tonnes (95,600 ounces) were sold to Developing Countries – Paraguay, Malaysia, and Uruguay. The total amount awarded to countries submitting non-competitive bids in the period June 1978 to May 1979 was 46.0 tonnes (1.48 million ounces). This quantity had to be deducted from the 388.8 tonnes (12.5 million ounces) which the IMF had proposed to sell in the period June 1978 to May 1980.

As a result, it was announced on 14th May, 1979 that, beginning in June 1979, the amount on offer at public auctions would be reduced to 13.81 tonnes (444,000 ounces) and that sales at this monthly rate would continue until May 1980. The IMF conducted its final auction on May 7th, 1980. Awards to bidders in the May 7th sale totalled 443,200 ounces of fine gold – 800 ounces less than the amount for which bids were invited – and were made at prices bid which ranged from $500.20 an ounce to $511.15 an ounce and averaged $504.90 an ounce. The remaining 800 ounces could not be awarded because they did not reach the minimum award of 1,200 ounces under the terms and conditions. Bids were received for a total of 1,822,000 ounces.

In all the Fund held 45 public auctions in which it awarded 23.52 million ounces (731.6 tonnes) to 51 bidders on competitive bids, and 1.48 million ounces (46.0 tonnes) to 13 monetary authorities on non-competitive bids. The proceeds of these gold sales amounted to $5.7 billion, of which $1.1 billion represented the capital value equivalent to SDR 35 an ounce that was added to the Fund's general resources, and $4.6 billion represented profits that were channeled to the Trust Fund for the benefit of developing member countries. Of the total profits $1.29 billion are being transferred directly to 104 developing countries in proportion to their quotas on August 31 1975. The remainder of the profits are available for loans by the Trust Fund to eligible developing members that adopt adjustment programmes to strengthen their balance of payments positions. By May 1980 Trust Fund loans equivalent to SDR 1.9 billion

Table 5.16. Fund Gold Auctions: June 2, 1976–May 7, 1980.

Date (1)	Place of Delivery (2)	Ounces Bid (thousands) (3)	Ounces Awarded (thousands) (4)	Subscription Ratio[1] (5)	Ounces Awarded to Non-Competitive Bidders (thousands) (6)	Number of Bidders Total (7)	Number of Bidders Successful (8)	Number of Bids Total (9)	Number of Bids Successful (10)	No. of Non-competitive Bids (11)	Price Range of Successful Bids (US $ per fine ounce) (12)	Average Award Price (13)	Profits (In millions of US dollars) (14)
1976													
June 2	New York	2,320.0	780.0	2.97	—	30	20	220	59	—	126.00–134.00	126.00	67.10
July 14	New York	2,114.0	780.0	2.71	—	23	17	196	56	—	122.05–126.50	122.05	64.00
Sept. 15	New York	3,662.4	780.0	4.70	—	23	14	380	41	—	108.76–114.00	109.40	53.82
Oct. 27	New York	4,214.4	779.6	5.40	—	24	16	383	37	—	116.77–119.05	117.71	60.25
Dec. 8	London	4,307.2	780.0	5.52	—	25	13	265	33	—	137.00–150.00	137.00	75.35
1977													
Jan. 26	New York	2,003.2	780.0	2.57	—	21	15	192	49	—	133.26–142.00	133.26	72.50
Mar. 2	New York	1,632.8	524.8	3.11	—	21	7	187	14	—	145.55–148.00	146.51	55.60
Apr. 6	New York	1,278.0	524.8	2.43	—	18	11	136	22	—	148.55–151.00	149.18	57.02
May 4	New York	1,316.4	524.8	2.51	—	17	14	107	38	—	147.33–150.26	148.02	56.37
June 1	New York	1,014.0	524.8	1.93	—	14	13	75	35	—	143.32–150.00	143.32	53.87
July 6	Paris	1,358.4	524.8	2.59	—	15	15	83	35	—	140.26–145.00	140.26	52.16
Aug. 3	London	1,439.2	524.8	2.74	—	18	16	136	44	—	146.26–150.00	146.26	55.31
Sept. 7	New York	1,084.4	524.8	2.07	—	15	11	115	21	—	147.61–149.65	147.78	56.24
Oct. 5	New York	971.2	524.8	1.85	—	17	12	103	32	—	154.99–157.05	155.14	59.97
Nov. 2	London	1,356.4	524.8	2.58	—	18	7	90	21	—	161.76–163.27	161.86	63.29
Dec. 7	New York	1,133.6	524.8	2.16	—	19	19	108	58	—	160.03–165.00	160.03	62.13
1978													
Jan. 4	New York	984.8	524.8	1.88	—	19	19	103	64	—	171.26–180.00	171.26	67.68
Feb. 1	Paris	598.4	524.8	1.14	—	17	17	76	62	—	175.00–181.25	175.00	69.65
Mar. 1	New York	1,418.0	524.8	2.70	—	19	16	127	76	—	181.13–185.76	181.95	72.92
Apr. 5	New York	1,367.6	524.8	2.60	—	21	15	122	30	—	177.61–180.26	177.92	70.78
May 3	London	3,104.0	524.8	5.91	—	24	17	192	36	—	170.11–171.50	170.40	66.83
June 7	New York	1,072.4	470.0	2.28	925.2	21	15	137	28	5	182.86–183.92	183.09	195.64

Table 5.16. (continued)

(1)	(2)	(3)	(4)	(5)	(6)	(7)	(8)	(9)	(10)	(11)	(12)	(13)	(14)
July 5	New York	797.2	470.0	1.70	20.8	22	19	101	44	2	183.97–185.01	184.14	68.96
Aug. 2	New York	1,467.6	470.0	3.12	70.0	21	20	117	42	2	203.03–205.11	203.28	85.84
Sept. 6	New York	773.2	470.0	1.65	133.6	20	10	89	25	2	212.39–213.51	212.50	101.42
Oct. 4	London	805.6	470.0	1.71	134.4	18	12	76	25	1	223.57–224.62	223.68	107.74
Nov. 1	New York	689.6	470.0	1.47	80.0	14	7	50	24	1	223.03–230.00	224.02	98.37
Dec. 6	Paris	1,965.2	470.0	4.18	20.0	16	13	102	31	1	195.51–196.75	196.06	74.23
1979													
Jan. 3	New York	1,479.6	470.0	3.15	16.4	17	9	159	23	1	219.13–221.00	219.34	84.73
Feb. 7	New York	1,489.6	470.0	3.17	59.2	19	5	123	11	1	252.47–252.77	252.53	109.60
Mar. 7	London	1,534.4	470.0	3.26	—	18	17	127	50	—	241.28–243.26	241.68	92.62
Apr. 4	New York	1,186.8	470.0	2.53	—	17	14	107	44	—	238.71–240.27	239.21	91.37
May 2	New York	1,514.8	470.0	3.22	20.0	20	17	155	56	1	245.86–247.01	246.18	98.79
June 6	New York	1,452.4	444.0	3.27	—	19	5	109	19	—	280.22–281.37	280.39	104.73
July 3	New York	1,518.8	444.0	3.42	—	20	13	113	23	—	281.06–281.87	281.52	104.84
Aug. 1	New York	1,138.8	444.0	2.56	—	20	16	133	63	—	288.95–291.07	289.59	107.84
Sept. 5	Paris	1,646.0	444.0	3.71	—	21	4	81	6	—	332.01–333.50	333.24	127.73
Oct. 10	New York	665.6	444.0	1.50	—	16	9	52	15	—	412.51–420.80	412.78	163.20
Nov. 7	New York	1,798.4	444.0	4.05	—	16	13	189	53	—	391.77–389.01	393.55	154.26
Dec. 5	London	1,746.0	444.0	3.93	—	18	15	97	38	—	425.40–429.31	426.37	169.27
1980													
Jan. 2	New York	1,342.4	444.0	3.02	—	10	5	52	10	—	561.00–564.01	562.85	229.17
Feb. 6	New York	1,939.6	444.0	4.37	—	17	5	80	8	—	711.99–718.01	712.12	295.24
Mar. 5	New York	1,412.4	444.0	3.18	—	16	14	84	54	—	636.16–649.07	641.23	263.52
Apr. 2	New York	802.8	444.0	1.81	—	16	16	69	66	—	460.00–503.51	484.01	194.97
May 7	New York	1,822.0	443.2	4.10	—	21	21	225	67	—	500.20–511.15	504.90	203.51
Total²			23,517.5		1,480.3								4,640.44

¹ The ratio of total bids to the amount on auction.

² Ounces awarded do not add up exactly to the total representing the amount sold owing to variations in the weight of standard gold bars.

had been made to 50 developing country members and an additional amount of approximately SDR 1 billion is available for the same purpose. Details of the total amounts of gold sold to individual members over the period are shown in Table 5.17.

The plans for the disposal of one third of the IMF's gold were made at a time when the decision makers felt confident that the SDR would eventually become the principal international reserve asset. Political, financial, social and industrial developments since the mid-1970s have resulted in a change in the priorities of those concerned with the gradual evolution of the international monetary system. The earlier problems associated with a lack of international liquidity, which had prompted the development of the SDR, have been replaced by difficulties caused by excess liquidity resulting from US deficits and the increased value of the reserves of those countries who valued their gold holdings at market related prices.

Consolidated Gold Fields Limited estimate that 544 tonnes were supplied to the non-communist private sector during 1979 as a result of reported sales by the IMF and USA together with their estimate of sales and purchases they considered were the result of government decisions. This represented some 32 per cent of the 1979 new gold supply. These official gold sales disappeared altogether in 1980.

Developments in the Gold Price

Factors influencing the gold price are many and complex. They include political, monetary, economic and even social conditions. From the gold mining industry's point of view it is imperative that the gold price should at least rise in real terms by somewhat more than the global rate of inflation to ensure reasonable profitability. Some of the factors affecting the gold price are listed, in no specific order of importance, below. They include: the freedom of central banks to buy and sell gold on the market, US Treasury and IMF auctions, the continuing industrial demand for gold, the investment and speculative demand for gold arising from economic and political uncertainty as well as currency instability worldwide, the diversification out of the dollar into other forms of investment, the growth in the popularity of gold coins, the rise in energy prices and the formation of the European Monetary System. J. Aron have prepared an excellent diagrammatic summary of the major factors affecting the gold price (*see Figure 5.5*).

The stability of the gold price from the Napoleonic era to the First World War, illustrated in Figure 5.6, remains, so far, a unique exception in world history and is explainable by the extraordinary development of national bank credit and paper money, on the one hand, and of nineteenth century gold production on the other. Gold production is estimated to have risen from a yearly average of about 0.37 million ounces in the

Table 5.17. Sales of Gold to Members in Four Distributions.

Member	Fine Ounces
Afghanistan	31,665.860
Algeria	111,258.000
Argentina	376,564.794
Australia	569,126.658
Austria	231,072.187
Bahamas	17,116.997
Bahrain	8,557.999
Bangladesh	106,978.879
Barbados	11,125.930
Belgium	556,289.958
Benin	11,123.669
Bolivia	31,665.985
Botswana	4,278.867
Brazil	376,564.569
Burma	51,349.512
Burundi	16,260.984
Cameroon	29,953.907
Canada	941,394.291
Central African Republic	11,125.995
Chad	11,125.762
Chile	135,220.043
China[1]	470.705.277
Colombia	134,364.514
Congo	11,126.000
Costa Rica	27,386.995
Cyprus	22,251.563
Denmark	222,515.702
Dominican Republic	36,800.979
Ecuador	28,241.989
Egypt	160,894.508
El Salvador	29,953.988
Equatorial Guinea	6,691.724
Ethiopia	23,106.991
Fiji	11,125.708
Finland	162,607.999
France	1,283,718.053
Gabon	12,836.997
Gambia, The	5,819.577
Germany, Fed. Rep. of	1,369,327.945
Ghana	74,456.884
Greece	118,104.738
Grenada	1,711.539
Guatemala	30,810.000
Guinea	20,497.774
Guyana	17,116.904

Table 5.17 *(continued)*

Member	Fine Ounces
Haiti	16,260.997
Honduras	21,395.994
Iceland	19,683.995
India	804,429.402
Indonesia	225,515.983
Iran	164,311.978
Iraq	93,273.618
Ireland	103,554.980
Israel	111,257.225
Italy	855,829.551
Ivory Coast	44,483.056
Jamaica	45,358.961
Japan	1,026,995.292
Jordan	19,683.976
Kenya	41,079.961
Korea	68,465.912
Kuwait	55,628.986
Lao People's Democratic Republic	11,125.533
Lebanon	7,701.998
Lesotho	4,278.316
Liberia	24,818.989
Libyan Arab Jamahiriya	20,505.334
Luxembourg	17,116.999
Madagascar	22,252.000
Malawi	12,836.998
Malaysia	159,165.213
Mali	18,826.368
Malta	13,692.991
Mauritania	11,124.698
Mauritius	18,827.858
Mexico	316,656.950
Morocco	96,625.279
Nepal	10,611.998
Netherlands	599,080.944
New Zealand	172,877.650
Nicaragua	23,105.366
Niger	11,124.483
Nigeria	115,536.844
Norway	205,398.687
Oman	5,990.856
Pakistan	201,097.447
Panama	30,807.853
Papua New Guinea	17,116.813
Paraguay	16,260.994
Peru	105,266.981

Plate 1. A scene from the Californian Gold Rush showing miners holding pans containing gold nuggets.
(Reproduced by courtesy of Gold Information Office)

Plate 2. Gold-bearing ore.
(Reproduced by courtesy of Consolidated Gold Fields Limited)

Plate 3. The electrolytic process produces gold of purity 99.5 per cent or higher. Dore gold is used as the semi-circular anode (left) which has been partially dissolved in a gold chloride solution. Pure gold has been deposited on the cathodes (right).
(Reproduced by courtesy of Gold Information Office)

Plate 4. Gold being poured into bars on a mine. The bars contain, on average, 88 per cent gold, 10 per cent silver and 2 per cent base metals.
(Reproduced by courtesy of Gold Information Office)

Plate 5. Refined gold bars stacked in the vaults of the South African Reserve Bank. Each gold bar has been weighed, checked for blemishes and stamped with a serial number and its fineness in parts per ten thousand, e.g. 9964.

(Reproduced by courtesy of Gold Information Office)

Plate 6. Gold bars, with a small proportion of copper bars, being deposited in a furnace. Krugerrands will eventually be minted from this gold alloy.
(Reproduced by courtesy of Gold Information Office)

Plate 7. One-ounce Krugerrands which are the world's best-selling gold coins. A large part of South Africa's newly minted gold goes into the manufacture of Krugerrands.
(Reproduced by courtesy of Gold Information Office)

Plate 8. The gold face mask of King Tutankhamun which is over 3 300 years old. (Reproduced by courtesy of Gold Information Office)

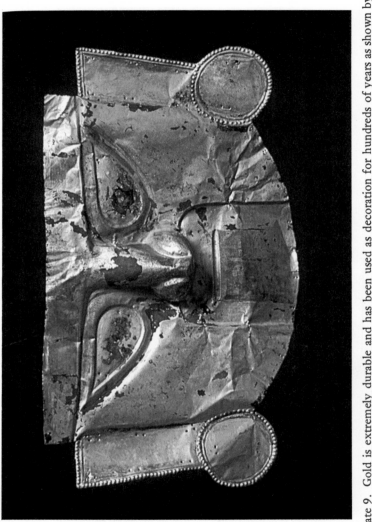

Plate 9. Gold is extremely durable and has been used as decoration for hundreds of years as shown by this mummy mask. It was made by the people who lived in Peru before the 15th century yet its radiance is untouched by the intervening 500 years.

(Reproduced by courtesy of Gold Information Office)

Plates 10 and 11. Most of the gold produced today is used in jewellery, and gold has always been used for decoration. Plate 10 (left) shows an 18 carat gold honeycomb with bee and fire opal honey droplets designed by John Donald. Plate 11 (right) shows an 18-carat choker and bangle designed by Mario Saba.

(Reproduced by courtesy of Gold Information Office)

Plate 12. The use of gold has also moved with the times as this NASA photograph shows. The moon landing vehicle has been coated with a thin reflecting layer of gold for protection from the sun. (Reproduced by courtesy of Gold Information Office)

Table 5.17 (continued)

Member	Fine Ounces
Philippines	132,653.664
Portugal	100,131.991
Qatar	17,116.942
Romania	162,589.303
Rwanda	16,260.959
Saudi Arabia	114,680.688
Senegal	29,024.482
Sierra Leone	21,395.847
Singapore	31,665.603
Somalia	16,260.991
South Africa	273,865.222
Spain	338,052.863
Sri Lanka	83,870.981
Sudan	61,619.962
Swaziland	6,846.993
Sweden	278,144.648
Syrian Arab Republic	42,791.963
Tanzania	35,944.820
Thailand	114,680.978
Togo	12,834.765
Trinidad and Tobago	53,916.983
Tunisia	41,066.775
Turkey	129,229.953
Uganda	34,232.819
United Arab Emirates	12,836.804
United Kingdom	2,396,322.177
United States	5,734,062.882
Upper Volta	11,124.368
Uruguay	59,051.913
Venezuela	282,422.976
Viet Nam	53,044.523
Western Samoa	1,711.788
Yemen Arab Republic	8,557.999
Yemen, People's Democratic Republic of	24,819.000
Yugoslavia	177,144.008
Zaire	96,708.971
Zambia	65,042.532
Total[2]	24,977,768.637

[1] Distribution was effected before April 17, 1980, when the Fund decided that the Government of the People's Republic of China represents China in the Fund.

[2] Arrangements have not yet been completed for sale of gold to Democratic Kampuchea. The notional share of this member in 25 million ounces is 21,396 fine ounces.

Source: International Monetary Fund Annual Report 1980.

previous three centuries, to 3.72 million ounces, or ten times as much, in the nineteenth century, and 19.23 million ounces in the thirteen years preceeding the First World War. According to Professor Triffin, paper money and bank deposits accounted for less than one third of the total money stock of the major countries early in the nineteenth century but by 1913 it accounted for 90 per cent of the world's money. Thus increased supplies of gold combined with a rise in the demand for liquidity being partially supplied by paper money replacing coins resulted in the gold price remaining stable during this time period.

As discussed in Chapter 3 the gold price was fixed from 1924 until

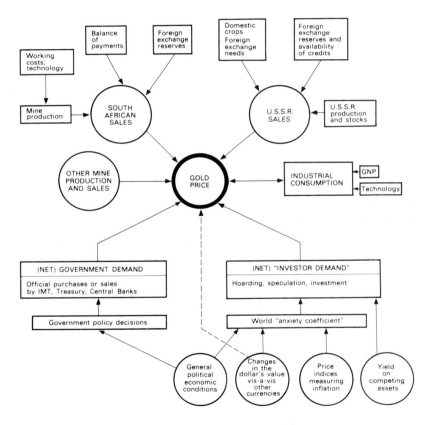

Chart prepared by J. Aron Precious Metals Research Department

Figure 5.5. Price-making influences in gold (arrows indicate principal causal direction).

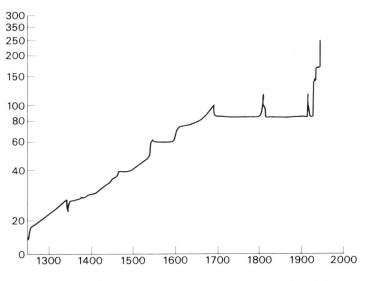

Source: Bank for International Settlements Annual Report 1951

Figure 5.6. Gold price in London over seven centuries.

1934 and again remained fixed between 1934 and 1968. During this latter time period the price was effectively pegged by the US Government at $35 per ounce. The period since then has been characterised by considerable price volatility. In spite of the volatility of gold prices, the overall trend has been strongly upwards since early in the decade. This volatility can be illustrated by gold price movements in the 1970s (*see Figure 5.7*). At the peak of the gold boom on January 21, 1980 no less than $850 was paid for one ounce of fine gold, compared with an average of 'only' $239 in April 1979 and $110 in the summer of 1976. Chapter 6 discusses the role of gold as an investment and illustrates some of the implications of this gold price volatility.

In contrast to the earlier periods of speculation, the gold boom discussed further in Chapter 6 was not accompanied in its final phase (i.e. roughly between the end of 1979 and the beginning of 1980) by a pronounced weakness of the dollar. In the wake of the Afghanistan crisis, investors apparently thought that almost all national currencies involved greater risks than before. The oil-rich countries, in particular, seem to have stepped up their demand for gold during that period in connection with their higher receipts of foreign exchange due to the oil price rises. The freezing of Iranian assets with US banks was no doubt another contributory factor. The supply of gold increased considerably in 1979 as, on top

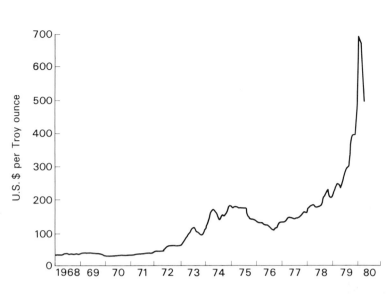

Source: *Gold 1980.* Published by Consolidated Gold Fields Limited

Figure 5.7. The London gold price.

of South African production and sales by the USSR, roughly some 544 tonnes of gold were sold by the US monetary authorities and the IMF from their holdings. However, these quantities were not large enough to meet the heavier demand for gold for industrial production and above all for private hoarding; the price of gold was therefore bound to soar. At the same time, the rising gold price greatly stimulated speculative demand. From the start, the excessive gold speculation involved the risk of a severe set-back, and between the end of January and the beginning of May, 1980 the price of gold had dropped by $340 per ounce; even so, at roughly $510 it was still twice as high as before the earlier wave of speculation which started in the Spring of 1979. By the end of 1980 the price had risen to around $620 per ounce but by August 1981 the price had fallen to $397 per ounce (this is discussed further in Chapter 6).

References

(1) Lipschitz and Otani, *IMF Staff Papers*, March 1977.
(2) Op. cit.

Chapter Six
GOLD AS AN INVESTMENT

Introduction

When deciding upon the structure of an investment portfolio an investor needs to decide in which investment areas he is interested and he must then decide which specific form his investment, in that area, will take. This Chapter examines both these questions. First of all the major investment criteria which now characterise the gold market are illustrated. Second, the pros and cons of investing in gold bullion, bullion accounts, bullion certificates, gold coins, gold shares, gold call options and gold futures are examined. Investment in gold jewellery is not discussed for two reasons. First, gold jewellery is often (but not always) bought for aesthetic reasons, and second there is traditionally a large 'mark up', often over 300 per cent, on this form of gold investment.

At the outset, the investor should assess the relevant percentage of his/her investments/assets which he/she wishes to commit to gold and gold related investments. Remember, to some degree or another, all these investments will be going up or down with the price of gold. The alternatives of holding cash, other equities or even other real assets such as antiques should all be borne in mind. There is no point in putting all your eggs in one basket and there is an opportunity cost, namely the interest which could be earned by putting cash into a deposit account, when investing in gold bullion. As important as making sure that not all one's eggs are in one basket is to ensure that a reasonable spread of gold related investments is achieved.

The purchasing power of gold over time has intrigued many economists including Professor Roy W. Jastram of the University of California. He has examined the history of general price inflation and the history of the price of gold in both England and the United States. This extensive study covers the time period 1560-1976 for England and 1800-1976 for the United States. Jastram's book, the *Golden Constant 1560-1976* demonstrates the close relationship between gold and commodity prices over the past four hundred years.[1] He shows that a brick or loaf of bread costs very much the same in terms of gold in 1960 as it did in 1560. Naturally within this long term steadiness there have been cycles of inflation and deflation when the ratio of gold to commodity prices has varied. Surprisingly he shows that gold tends to lose its purchasing power during inflationary periods and to gain purchasing power during deflationary periods. Yet, even though gold has been a poor hedge against year to year commodity price increases, it has maintained its long term purchasing power from inflationary peak to inflationary peak. That is, commodity prices repeatedly return to the same price in terms of ounces of gold, what Jastram calls the 'retrieval phenomenon', an observation that cannot be confirmed for the prices of manufactured goods, which reflect steadily rising labour costs. Gold can thus function as an effective long term hedge not only against inflation but also as a short term hedge against deflation because its buying power appreciates more rapidly than anything else during such periods.

The publication *Gold 1980* issued by Consolidated Gold Fields Limited sets out three major characteristics that an investment area will fulfill when it has received the full attention of long term investors. These are:

(1) It will possess a large number of individuals and financial organisations who have confidence that it will give a satisfactory performance compared with other investments and hopefully will keep pace with or, better still, outpace inflation.

(2) Periodically, it will become the recipient of the massive investment funds which can be mobilised rapidly by the international financial community.

(3) New participants will be drawn into the market during periods of rising prices as a result of increased attention from the press, radio and television.

The evidence put forward by Consolidated Gold Fields Limited that gold satisfies the first criteria is that during the 1970s a wide range of methods for investing in gold increased in popularity. These included the North America futures markets, bullion accounts, gold certificates and sales of coins and bars. Gold coins, particularly the Krugerrand, have been very successful and the total coin market rose from 46 tons in 1970 to 290 tons in 1979. *Gold 1980* goes on to state:

'The main reason for the widening of the market for gold is the lesson of the 1970s that gold, despite price fluctuation, can be used as a means of maintaining the real value of savings. Secondly, Figure 6.1 demonstrates that the gold price has shown it can increase against all the major currencies. For several years there was no significant improvement in terms of the Swiss Franc, German Deutschemark and the Japanese Yen but they have eventually had to yield to gold. An additional factor is that it is now unlikely that very large sales will be made from the "above-ground" stocks of gold which are held by Central Banks, International Monetary Fund, European Monetary System and the private sector. Although some sales will be made it is felt that the market can absorb them without any large price reductions. Ten years ago, many investment managers believed that it was possible to continue to have high growth rates, low inflation rates and an international monetary system which did not require gold. Under these circumstances it was argued that the Central Banks would sell these gold holdings of some 35,000 tonnes and keep the gold price at low levels for a very long period. The increases in the price of oil and the subsequent high inflation rates together with low growth rates have removed the fear of massive sales of "above-ground" stocks of gold.'

Evidence for the appreciation of gold in terms of all major currencies is also provided by Table 6.1. These statistics provide convincing evidence of the role that gold has played over the last ten years as a store of value in a world increasingly distrustful of paper currencies.

In order to appreciate why gold periodically becomes the recipient of massive investment funds Consolidated Gold Fields stress the necessity of appreciating the investment cycles which occur as economies move from recession to expansion. In an attempt to reduce inflationary expectations western governments have been cutting back monetary growth, thereby forcing up interest rates. During this time period investors find it profitable to place funds in bank deposits rather than gold. The problems associated with holding a non-interest bearing real asset at a time when real rates of return have been available on deposit funds were clearly illustrated by the decline in the gold price over the first six months of 1981. At the same time as funds are flowing into bank deposits other investments such as ordinary shares and bonds fall until interest rates reach the level necessary to slow down business activity thereby breaking the inflationary expectations. As the recession deepens the demand for loanable funds diminishes and interest rates start to fall. Under these circumstances bonds, and all other fixed interest securities, will be in heavy demand since their prices rise automatically as interest rates fall. As unemployment rises, the political democracies are forced to reflate and as inflation eventually takes off funds will quickly flow into the gold markets. With the large investment income alone of OPEC, discussed in detail in Chapter 8, combined with the new investment cash flows generated in North America, Western Europe and Asia, there then arises the potential for rapid gold price increases.

The reason why new participants are drawn into the gold investment area, Consolidated Gold Fields Limited go on to illustrate, is that initially when the demand for gold is particularly strong, supplies from new mine production are quickly taken up, forcing up the price quite considerably before private stocks of gold come onto the market. During the major

Table 6.1. Gold price changes.

GOLD PRICES	Price @ 5 May 1971	GOLD PRICE			
		5 May '71 and Aug '71	Aug '71 ¢ and 19 Mar '73*	19 Mar. '73 and End year '73	Beginning and End '74
US dollar	39.86	+ 7.8	+ 91.2	+ 36.4	+ 65.7
Deutsche mark	145.88	− 5.1	+ 58.8	+ 30.7	+ 48.5
French franc	221.30	− 0.7	+ 71.0	+ 38.7	+ 57.2
Belgian franc	1993.00	− 3.3	+ 66.9	+ 39.7	+ 45.0
Dutch guilder	144.29	− 3.3	+ 66.9	+ 31.8	+ 47.2
Danish krone	298.95	+ 0.4	+ 73.4	+ 36.1	+ 49.0
Norwegian krone	284.71	+ 0.4	+ 72.7	+ 30.3	+ 51.1
Swedish krone	206.20	+ 0.4	+ 81.9	+ 37.0	+ 47.2
Austrian schilling	1036.36	− 3.3	+ 62.7	+ 31.5	+ 43.3
Pound sterling	16.61	+ 6.5	+ 89.2	+ 45.0	+ 63.9
Italian lira	24912.50	+ 0.4	+ 89.3	+ 44.9	+ 76.6
Swiss franc	174.30	− 5.3	+ 53.4	+ 36.0	+ 29.2
Japanese yen	14349.60	− 7.3	+ 51.2	+ 44.8	+ 78.2
S. African rand	26.75	+ 7.8	+ 106.3	+ 36.4	+ 75.2

x US dollars per pound
¢ Smithsonian realignment   *Currency crisis – European markets closed
Source: Laurence Prust & Co.

price rises in the 1970s the attendant media publicity brought in new buyers whose first priority was to obtain gold and who appeared not to have made a close study of price trends before making purchases. As the price declines many of the new entrants, despite potentially having bad experiences, will stay with gold and adopt the characteristics of the longer term more experienced goldholders.

Consolidated Gold Fields Limited stress that the above supply and demand forces are inadequate when it comes to explaining the dramatic price increase to $850 in January 1980 and the dramatic fall to $474 in late March. The extra force driving up the gold price was heavy buying of gold from the Middle East in response to the freezing of the Iranian assets by the USA and also as a consequence of the fear for personal safety as a result of the civil disturbances in Mecca, the movements of large numbers of Russian troops in Afghanistan, the possibility of political problems in Yugoslavia and the fear that the USA would take military action to free the hostages in Teheran. The subsequent price fall, partially reversed on the news of the outbreak of war between Iran and Iraq, was caused by the mobilisation of large sales from privately held stocks of gold.

The advantages and disadvantages of investment in gold depend on an individual's perception of the circumstances at the time, future events and government policies. Whilst gold has many features in common with investment in shares, properties or insurance policies, it also has some

Table 6.1. *(continuation).*

CHANGES BETWEEN

Beginning and End '75	Beginning and End '76	Beginning and End '77	Beginning and End '78	Beginning and End '79	Beginning Jan and 2 Apr 80	Cumulative Change 2 Apr 80
− 24.4	− 4.2	+ 18.5	+ 37.1	+ 125.29	− 1.38	+ 1161.9
− 18.0	− 13.8	+ 8.5	+ 18.9	+ 113.17	+ 12.73	+ 579.5
− 24.0	+ 6.4	+ 13.6	+ 22.	+ 117.05	+ 11.28	+ 931.6
− 17.4	− 13.0	+ 10.5	+ 21.2	+ 118.40	+ 11.1	+ 675.7
− 19.1	− 12.3	+ 11.7	+ 19.1	+ 117.00	+ 11.73	+ 649.9
− 17.3	− 10.3	+ 18.3	+ 20.9	+ 137.96	+ 11.97	+ 926.5
− 19.1	− 13.0	+ 19.5	+ 33.9	+ 121.56	+ 0.45	+ 822.5
− 18.3	− 10.3	+ 27.3	+ 27.3	+ 117.99	+ 0.75	+ 1003.6
− 18.4	− 13.1	+ 9.7	+ 21.1	+ 109.32	+ 12.03	+ 584.6
− 12.4	+ 14.1	+ 8.1	+ 28.7	† 105.82	+ 0.29	+ 1313.6
− 20.2	+ 22.6	+ 18.1	+ 30.7	+ 118.33	+ 11.76	+ 1740.4
− 21.7	− 11.6	− 0.2	+ 11.4	+ 121.87	+ 15.73	+ 439.6
− 23.4	− 7.8	+ 0.2	+ 11.2	+ 178.28	+ 0.44	+ 791.8
− 7.5	− 4.3	+ 18.5	+ 37.1	+ 113.99	− 3.31	+ 1422.6

differences. For example, gold bullion does not generate interest or dividend income. Certain types of gold investment, namely the purchase of gold bullion and bullion coins, require much less professional expertise than do decisions about the purchase of gold shares, gold futures and call options. Consequently, this Chapter gives up much more time to discussing the latter forms of investment than the former. This is not to say that these are more important or profitable forms of gold investment, simply that conceptually they require more professional expertise than the former. Table 6.2 outlines the characteristics of alternative forms of gold investment.

Gold Bullion, Bullion Certificates, Bullion Accounts and Delivery Orders

Gold Bullion

Gold bullion in two grades of purity is cast into bars with a wide range of weights and dimensions for investment purposes. The first grade of purity is that which is characteristic of bars conforming to the specifications of the London Gold Market for 'good delivery' bars. These must contain a minimum of 99.5 per cent Au (995 parts per 1000 fine gold), conform to certain physical specifications and be marked by a serial number and the stamp of an 'acceptable melter or assayer'. If not marked with the fineness and stamp of an acceptable assayer each bar must be accompanied by

Table 6.2. Characteristics of alternative forms of gold investments.

Object	Premium over dealer's gold price	Authenticity	Storage*	Liquidity	Other
				Definition: ability to buy or sell without having to accept a major premium or discount when doing so	
Physical 1 Gold bullion in bars of: a 1 kg b 100 oz c 400 oz	The larger the bar, the more acceptable the brand and the closer the quantity sold is to 2,000 oz, the lower the premium. These bars thus have a lower premium than ½-oz or 1-oz bars or bullion coins	Problematic unless the bars bear weight and fineness markings and the stamp of a well-known and widely accepted refiner or bullion dealer	Necessary. Can be accomplished at home, in safe, in safe deposit box, or at recognized reputable	Best liquidity is in quantities of 2,000 to 4,000 oz. Lower quantities bear a transaction cost which may slightly (perhaps $50 to $60 transaction) impinge on liquidity; higher quantities placed on the market at one time may, if the market is sensitive, affect it	The smaller the bar the easier to store at home or to transport

Table 6.2. (continuation).

Object	Premium over dealer's gold price	Authenticity	Storage*	Liquidity	Other
2 Gold Coins Bullion: a Mexican 20-peso b Mexican 50-peso c Aust. 100-kroner d South African Krugerrand	Although bullion coins have a slightly higher premium over their gold value than bullion bars, they may have as low a premium as small bars and may retain some or all of this premium value on re-sale		warehouse against issuance of a receipt (cost of latter – with insurance – approximately ½% p.a.)	There is a slightly wider spread between the bid and asked prices of coin than between those of bullion, but the market for the well-known bullion or trade coins listed at the left is nevertheless highly liquid	Since coins come in very small units – down to ½ oz for the Mexican 20-peso coin – they have the highest measure of divisibility, that is ability to be sold in small units. Even in small units, the seller may not experience an excessively adverse price because of their easy authenticability and the breadth of the coin dealer market
Trade: e British Sovereign f US Double Eagle	Trade coins clearly have a higher premium and will almost surely retain a rarity premium	Counterfeiting is possible but can usually be detected by a coin specialist even without destructive assay			

Table 6.2. (*continuation*).

Object	Premium over dealer's gold price	Authenticity	Storage*	Liquidity	Other
Non-physical 3 Gold futures	The gold futures market has the lowest premium of all because it is actively trading. The buyer cannot, however, know the momentary market price without close contact with the trading floor. The premium is the difference between the prevailing bid and asked price plus brokerage commissions and the interest and liquidity cost of the original margin used to secure the obligation	Usually unnecessary to check as weight and fineness are guaranteed by chain of previous owners, including the clearing member who first placed the bullion into the vault and the refiner from whom that clearing member received the bullion	Buyer need not be concerned about storage until he takes delivery; his status is then no different than he who owns physical gold bullion as in 1 above. Once he takes delivery, he must make sure that the material is insured and that in case of loss the insurance company will pay the holder of the warehouse receipt directly	High liquidity except on 'limit move' days, that is days when the market has moved so far up or down that exchange rules mandate closing the market until the following business day	Gold futures provide the buyer with the greatest amount of leverage, but by so doing exposes him to the risk of variation margin calls which may cause him to be closed out. In addition, the buyer is dependent on the continued solvency of the commodity broker with whom his order is placed

Table 6.2. (*continuation*).

Object	Premium over dealer's gold price	Authenticity	Storage*	Liquidity	Other
4 Gold bullion for delivery at bullion broker's vault in London	No premium over gold price except ¼% commission if bought at 'fixing'	No concern is necessary as authenticity is guaranteed by bullion broker (all of whom are subsidiaries of major banks or commercial companies)	Bullion will remain in broker's vault until withdrawn or sold. Approximate cost – which includes insurance – is ½% premium	Total liquidity Resale is without commission. High quantities may be dealt on fixing without impact on price	Oldest and best-known market in the world. 'London good delivery' bars and vaults are the world's standard. Highly convenient method of holding gold
5 Delivery orders for gold bullion or coin	Modest premium to reflect cost of issuance	No concern is necessary since issuer stands behind bullion described on face of delivery order	Specific material for the buyer is sold whilst in storage at recognized precious metals warehouse of buyer's choice (Iron Mountain Depository Corporation in New York, First National Bank of Chicago in Chicago or MAT in Zurich). Insurance payable to holder of delivery order is covered in cost of storage. Cost of storage and insurance approximates ½%	As liquid as the underlying object. Fully acceptable for payment of gold obligations to Mocatta & Goldsmid Limited and other leading bullion dealers	Most convenient way to hold gold in readily transferable form

Table 6.2. (*continuation*)

Object	Premium over dealer's gold price	Authenticity	Storage*	Liquidity	Other
6 Gold bar shares or trust fund	Premium is equal to the 'load' of the fund plus commission paid by the fund to the sellers. Currently described funds have a 'load' between 5 and 10% with annual management fees of 0.35 to 1.0%.	No need for concern if fund managers buy bullion through reputable refiners or bullion dealers and material remains in control of first-rate institutions and is set aside by these institutions for fund only	Arranged in bulk by fund managers or trustees. Buyer does not have right to a specific bar of gold	May be resold to fund at net asset value as calculated from day-to-day or through fund's agent at that day's gold price, in either case with a 1 to 2% charge	
7 Gold mining mutual funds and gold mining company shares	Not directly related to gold price. Subject to management performance, labour conditions, political risks, government intervention, taxation, other investments of mining company, etc	Not applicable	Not applicable	Related to shares market	Have a premium over 'intrinsic' value because these are currently the only bullion-related investments available to Americans. Earnings and thus share prices may increase because of high leverage and because rising gold prices coupled with stable production costs cause earnings to multiply

*For the owner who may want subsequently to resell, it is important that the gold items he buys be stored in a recognized reputable facility from the time he buys them until he sells them as this assures the next purchaser of their authenticity and/or the ability to rely on the name of the organization that sold it to the individual from whom he buys it.

Source: Mocatta Metals Corporation.

a certificate, issued by an acceptable assayer, stating the serial number of the bar and its fineness. Lists of the names of acceptable melters and assayers are issued by the London Gold Market. The second grade of purity is that which is achieved by electrolytic refining from which bars containing up to 99.99 per cent Au can be produced according to requirements.

Investment in gold bullion, whether in the form of bars or wafers, provide a direct investment into gold, which is instantly convertible into cash everywhere and has an internationally quoted price. The major disadvantages of holding bullion are that capital is tied up in an asset which yields no dividends (although it does provide the potential for capital gain) and storage charges may be expensive. A further disadvantage for United Kingdom investors is that they have to pay value added tax (VAT). One way of escaping the VAT entanglement is simply to channel gold purchases via foreign centres where no sales tax is applied. Now United Kingdom exchange controls have been scrapped, foreign bank accounts can be used lawfully to purchase gold free of VAT in Zurich, Hong Kong, or Luxembourg. The major factors affecting the bullion price were outlined in detail in Chapter 5.

Bullion Certificates

An increasingly popular form of gold investment is the gold certificate. A gold certificate is a non-negotiable depository receipt evidencing ownership of a specific amount of gold bullion. For investors not wishing to accept the storage costs of holding gold bullion, gold certificates provide an interesting alternative. The issuing institution, for example Citibank, registers the bullion certificate in your name and hands it over to you. The certificates generally state that you have the right to demand at any time the actual gold you have purchased. Alternatively, you can cash the certificate in and obtain the market value of your investment. Thus the advantage of gold certificates is that the investor has no storage risk and he is provided with a liquid, direct investment into gold. The major disadvantage as with a direct bullion investment is that one's capital is tied up in an asset yielding no dividends.

Bullion Accounts

Some banks offer 'bullion accounts'. These are simply accounts denominated in ounces of gold with the bank at which you hold the accounts guaranteeing the gold for you. The major problem with these accounts is that the minimum deal may be quite high (100 ounces as compared with 10 ounces for certificates), involve capital being tied up and in addition one may not easily be able to achieve delivery of the gold if one so wishes.

Delivery Orders

These were developed by Mocatta Metals Corporation. A delivery order is a document that gives the owner a piece of paper embodying direct ownership of physical gold. It essentially states 'You own a bar of this size with this number unit at this warehouse'. The warehouses are in tax free locations like Wilmington, Delaware or MAT Securities AG in Zurich, where it is possible to own the actual object tax free. These bars are often syndicated i.e. shared between several owners.

The bullion dealer signs the Delivery Order to prove that he put it there; the warehouseman signs it to show that he received it and the insurance company stipulate that if the gold is missing they will pay the holder directly. Delivery Orders sell at a 2%–3% premium and there is a well organised method for transferring them from one person to another, for taking delivery and for paying for storage – all anonymously.

If syndicated there may be problems in buying or selling where there are other parties to be considered. In addition the anonymity may not be complete as there must be a means of proving the ownership of the delivery order upon its theft or destruction.

Gold Coins

Like gold shares, gold coin prices are only partially affected by gold bullion prices. The major factors affecting gold coin prices are the gold content, the numismatic value, the condition, government taxes and regulations. Coins can be divided into bullion coins and numismatic coins. There is no official definition of numismatic coins, but in practice they designate official coins of a country, struck with special care and other coins struck to commemorate events of historical or national importance. Numismatic coins derive their value from their beauty, rarity and physical condition rather than from their gold content. Bullion coins such as Krugerrands, Napoleons, and Sovereigns, are bought primarily for their gold content, although they may also have artistic and fashion appeal as jewellery items.

One of the most important considerations to take into account when investing in gold coins is the 'premium' on the price. This premium measures the difference between the gold contained in a coin and the selling price of a coin. The premium can be calculated by multiplying the gold price by the gold content of the coin, and then calculating the percentage mark up that one has to pay to buy the coin. Thus a French 20 franc Napoleon with a gold content of 0.187 ounces combined with a gold price of $550 will have a gold content value of $102.5. If the current Napoleon selling price is $140, the price difference $37.5, expressed as a percentage, gives a premium of 36.5 per cent. The lowest premiums are those on bullion coins. These coins have no numismatic value and consequently one does not require extensive knowledge apart from the current gold price in determining their value. The premiums during 1979 of some of the most traded coins are given in Table 6.3.

Today's bullion coins are produced in large quantities by countries

Table 6.3. Premium of gold coins over their gold content.

Expressed in per cent. 1979	2 January	30 March	29 June	28 September	31 December
Based on gold prices of	$226.80	$240.10	$277.50	$397.25	$524.00
Sovereign (Old)	18	32½	34½	27	20½
Sovereign (New)	12½	10	10	9½	9½
U.S.A. $20	34½	37	47½	27½	17½
French 20 Francs	46	46	42	37½	28
Swiss 20 Francs	49½	46	39½	43	28
German 20 Marks	53	42	21	13½	8½
Krugerrand	2½	2½	3½	2½	2

Source: Samuel Montagu.

guaranteeing their gold content. Some countries, such as Mexico, Hungary and Austria have decided to mint 'restrikes', which feature the design of a coin that once was used as legal tender. In other cases, such as South Africa, a new design is used or the more recent date of issue is printed onto the coin. Currently, Austria, Hungary, Mexico, USSR, South Africa, the United Kingdom and Canada issue bullion coins. The South African Krugerrand was first minted in 1967; since then, Krugerrand sales have passed 30 million. The Krugerrand's price throughout the world is based directly on the price of gold bullion and the Krugerrand is possibly today's most negotiable gold vehicle. Krugerrand and the Canadian Maple Leaf gold coin weigh exactly one troy ounce. As discussed on p. 117 lower denomination Krugerrands have recently been issued. US gold coins were discussed in Chapter 2.

Bullion coins are designed to provide the investor with a direct link with the gold price while providing some interest in that the investment is in the form of a gold coin. Numismatic coins are, however, governed by four other factors although the gold price will provide a minimum beneath which the coin will not fall. These four other factors are rarity, age, quantity originally produced and their condition.

Determining the 'grading' or condition on a numismatic coin is a difficult task. Below is a listing of the most important gradings:

1. *Proof*
   Coins struck especially in small numbers for investment, presentations and other purposes. These coins are usually struck twice ('double-struck') and as a result have a high, mirror-like finish. Due to the small number produced and the great demand among collectors, proof coins usually sell at a considerably higher price than the standard issue.

2. *Brilliant Uncirculated (B.U. or U & C)*
   These coins are struck on regular dies, althoug' they are not intended for circulation. Most bullion type coins and ost numismatic coins produced today remain in brilliant uncirculated condition. In other

words, their holders try deliberately not to damage the coins in order not to reduce their investment value.

3. *Extremely Fine (E.S.) or Extra Fine (X.S)*
   These coins are slightly circulated and show some signs of wear, but still have their original polish.

4. *Very Fine (V.F.)*
   Details and design are still clearly marked; only the highest surfaces are worn down slightly.

5. *Fine (F)*
   Coins which have been circulated and show definite signs of wear, especially on the finer parts of the image. However, all the details are visible. The grading 'fine' is the minimum standard for coin collectors.

6. *Very Good (V.G.)*
   The inscriptions and design are still clear and bold, but worn.

7. *Good (G)*
   Design and inscriptions are readable but quite worn.

8. *Fair (F)*
   Features are still identifiable but not easily readable.

9. *Mediocre (M)*
   Very worn or damaged.

10. *Poor (P)*
    Unless it is very rare, a coin in 'poor' condition usually does not fetch more than the value of the metal it contains.

In conclusion, the advantages of investing in gold coins are that they are easily convertible into cash, negotiable internationally and provide a direct investment into gold. The major disadvantage for both types of coins is that one's money is tied up in the investment (unlike investments in futures and options) and that one may, particularly with numismatic coins, have to pay a high premium over the bullion price. Storage costs may also be important. In addition, with numismatic coins, one requires some expertise to judge the value of your investment and these types of coins may not always be readily negotiable. The Appendix on p. 239 provides a listing of the world's most traded gold coins.

Gold Shares

The equity financing of the major gold discoveries during both the last and present century has led to an extensive global market in gold shares. The shares of gold mining companies offer an alternative to other types of investment and are compatible with the development of an overall portfolio strategy including coins, bullion/bullion certificates/gold futures contracts.

The main advantage of gold shares is that, if profitable, a dividend is paid to the holder whereas a direct investment in gold, involving storage and insurance costs, may result in a negative income return. Another alternative, gold futures, with its attendant higher risks, may not be to every investor's taste. The criteria to be used in evaluating the monetary yield from gold share investments are described in Appendix I (*p. 198*). Appendix II (*p. 201*) describes two unique forms of gold share investment, the gold backed bonds issued by the French government.

Mining shares differ from industrial equities because the business of mining is basically different from any other industrial process. A manufacturing business or a service industry can, theoretically, go on forever, so long as its product is in demand. Thus the share prices of companies in these industries can discount any amount of future profits and future growth, if the market permits it. Mining is not like that. The business of mining is extracting ore from the ground and treating it to produce a metal which can then be sold to a manufacturer. Every ore body is limited in size and, once it has been mined out, that mine is dead. The business ceases to exist when there is no more ore. The value of a share in a mine is therefore simply the value of the dividends which that mine will produce during its life. A very small number of mines have such vast ore reserves that their life expectation is unimportant (the Rustenburg platinum mine in South Africa, for example, has reserves that will last for at least fifty years); but in general it is important to regard a mining share as a wasting asset. Part of the income from a mining share should therefore be regarded as a capital repayment.

Another peculiarity of the mining industry is that the concept of growth hardly exists. It tends to be taken for granted that any manufacturing or service business, if successful, can be expanded almost without limit; that assumption can be built into share prices on occasions. If one expands a mine, however, one normally shortens its life because one extracts the ore faster. This will not be true, however, when the expansion includes the milling of low grade ore which was previously left in the ground at a time of lower gold prices.

Investment Strategy

Whilst this section is basically devoted to the factors affecting gold mining shares there are a number of ways of gaining this exposure and a few points of difference between the various types of gold share investment will be examined. In addition, the function of each type of investment within the investor's portfolio will be analysed. The mining of any ore, whether it be gold or copper, is the realisation (hopefully at a profit) of a wasting asset. Consequently the return on a 'straight' gold mining company is comprised of two separate elements: return *on* capital and return *of* capital. As already mentioned, part of the income from a mining share should be regarded as a capital repayment. Although mine life is therefore a finite limit, it is not precise as changes in the price/cost relationship can make hitherto uneconomic areas profitable and mine lives can be extended. This is an important consideration in selection of 'marginal' or 'break-up' category mining shares.

To reduce the problem of the finite reserve life, mining companies themselves have developed the mining finance house system (these are defined and discussed in detail on pp. 177, 178). Each mining house has a number of mines in its group at different stages of exploitation and therefore have infinite lives under proper management. Mining finance houses sometimes provide a spread of other interests but, due to their longer term security, usually offer a lower return than direct mining investments. Finally, a number of gold related investment trusts, mutual funds or unit trusts exist. These manage portfolios in gold shares for the investor and their success depends on the type of portfolio and expertise of the fund managers.

For every reward there is a concomitant risk. Therefore, a mining finance house which is partially removed from the 'direct risks' of mining hazards provides a lower return than investment in a single hole in the ground. In addition, there are risks associated with geographical location and equity investment or what is known as 'fundamental risk'. This is discussed later in this section (see p. 179). Since risk and reward are linked it is possible both to define a risk/reward profile and to gauge by the volatility in the price of the asset its 'gearing' to the gold price. Gearing can evolve for a number of reasons but put simply it means the degree to which the value of the investment rises or falls with the gold price. To categorise this volatility or risk it is possible to rank (approximately) the different types of gold investment as follows.

(1) Mining finance houses, investment unit trusts (mutual funds).
(2) Mining holding companies.
(3) Straight mining shares:
    (a) Low cost, long life mines.
    (b) Medium cost, medium life mines.
    (c) High cost, short life mines.
(4) Gold futures.

There is a graduation of total return running through the above list which would range from around 5 per cent to well in excess of 100 per cent. It is important that this is borne in mind when planning investments and an investor has the flexibility to alter his risk/reward profile infinitely depending on his assessment of the expected moves in the gold price. Put simply these strategies amount to:

Gold price falling — Hold only mining finance houses, investment trusts etc. Put cash on deposit.

Gold price static — Maintain proportionate spread of investments in roughly equal amounts.

Gold price rising — Buy high cost, short life mines and gold futures contracts. A rise in selling price benefits disproportionately mines which are operating on low margins. Such mines could typically be those with high extraction costs, a low grade ore body or a short life.

Currencies and Market Considerations

The main market for gold shares is in London which firmly regained its former premier position on the ending of exchange controls in October 1979. However, dealing is an international operation covering the world's major financial centres and the prices of gold shares may be expressed in US dollars, pence sterling or local currencies. In nearly all gold share dealings there will be an associated currency transaction and investors should bear in mind that they have an investment in the respective currency as well as in the share itself. This is important to note when comparing share price movements against the gold price which is priced in US dollars. In December 1979 a survey carried out by a firm of South African stockbrokers showed that the holdings of South African gold shares by overseas investors amounted to 38.7 per cent on a weighted average basis. American investors represented the major proportion of foreign investors accounting for 23.2 per cent. Non-resident shareholders on the London register constituted the next largest foreign holding with 12.5 per cent, followed by shareholders on the French and Belgian registers which, together, held about 4.6 per cent of the total number of South African gold shares in issue.

Following the government's acceptance of the interim report on exchange rates of the De Kock Commission of Inquiry into the country's monetary system and monetary policy, South Africa's adherence to a fixed exchange rate *vis-à-vis* the US dollar was officially abandoned in January 1979. Under the dual exchange rate system currently in use, a majority of foreign transactions is conducted through the so-called 'commercial' rand; although still determined largely by the Reserve Bank, the exchange rate of the commercial rand is revised frequently in response to market developments. The other exchange rate, the 'financial rand' is the medium through which all non-South African residents purchase South African gold shares irrespective of whether the price and settlement was effected in US dollars or pounds sterling.

Financial rands can be purchased both in South Africa and overseas. While supply and demand does influence the price of the currency, the price is largely a function of the price differential of South African shares on the Johannesburg Stock Exchange and the same quoted shares on overseas markets. The discount of the Financial rand to the Commercial rand, as can be seen from Figure 6.2, has varied widely in the last ten years from zero per cent to 46 per cent, reflecting international economic events as well as foreign investor confidence in the economic and political outlook for South Africa. Financial rands could never stand at a premium over Commercial rands because it would then be cheaper to invest in South Africa through the banking system at the normal rate of exchange.

South African residents may not export the commercial rand to purchase shares on overseas markets. Instead they must use this currency to purchase financial rand (previously known as the securities rand) which

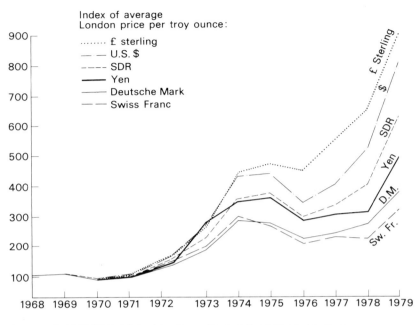

Source: *Gold 1980.* Published by Consolidated Gold Fields Limited

Figure 6.1. Gold prices in index form.

stands at a discount to the commercial rand. This means that South
African residents have to pay more for South African shares than overseas
residents (depending on the prevailing financial rand discount rate). Conversely, because the overseas investor can use the financial rand to purchase South African shares, he can obtain such shares at a discount to the
Johannesburg market quotation. The rate of the prevailing financial rand
discount depends on a number of factors, but basically it is a measure of
overseas (international) confidence towards investment in South Africa.
A rise in international confidence will cause the financial rand discount to
narrow, and a fall in this confidence will cause the discount to rise for non-
South African investors.

The benefit of being able to use financial rands is that such investments
could be made at a discount to the commercial rand rate and there is no
prohibition to disinvestment other than the discount on financial rands.
In addition, all dividends earned by non-residents are payable in commercial rands, therefore pushing up – while there is a discount on the financial
rand – the effective earnings yield on such investments. Thus the yield to
an overseas holder of South African securities is increased by the amount

of the prevailing financial rand discount (*see Figure 6.2*) at the time of the share acquisition.

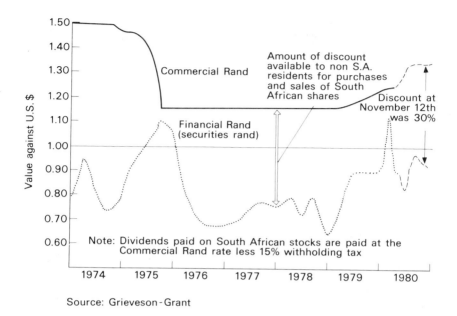

Figure 6.2. The commercial and financial rands in US dollars (1974–1980).

Consequently the price (and subsequent yield) of a South African gold share investment varies with investors' appraisal of future risks of investing in South Africa as well as the gold price.

South African shares can be bought on the Cape or London registers as registered securities, or in the American Depository Receipts (ADR) form for US investors. The latter category (ADRs) have become a popular and convenient means of holding shares in South African companies conforming with the North American bearer security form and are dealt in on the Over-the-Counter market (OTC). ADRs are available through four depository banks, Morgan Guaranty Trust, Citibank, Irving Trust and Chemical Bank, which mandate dividends on to the beneficial owners. There is a withholding tax of 15 per cent on all South African dividends which is not refundable but may be offset against an overseas tax liability where there is a Double Taxation Treaty. A similar withholding tax is levied on North American securities.

Dealing in Australian, Canadian and US gold mining companies is slightly less complicated and involves a direct investment in the currency concerned. This is because these companies do not operate a dual exchange rate system. Australian stocks may be dealt in London on the Australian or London registers but for settlement and delivery in Australia.

One way for US investors to gain an interest in South African gold shares is through the investment trust ASA (formerly American-South African Investment Company Limited). ASA is an unusual company in that it has South African registration and normally deals in shares only in Johannesburg, but in so far as South African residents are concerned, is classed as a foreign company in which they may not hold any of the shares without special permission. This goes back to the formation of the company in 1958, its main purpose being to enable American citizens in particular to invest in South African gold shares through a company listed on the New York Stock Exchange and with a relatively free market in the shares. The shares are also quoted in Amsterdam and dealt in on most European Stock Exchanges.

ASA is specifically exempted from South African taxation on profits arising from the realisation of its investments. In December 1974 the South African Reserve Bank confirmed that ASA is entitled to transfer to the USA in dollars, at the current rate of exchange ruling from time to time, both its capital and its gross income whether received as dividends or as profits on the sale of investments. This underlies the policy of calculating the net asset value on the basis of prices ruling in Johannesburg. ASA warns potential investors that the implementation of this policy by the South African Reserve Bank 'could be influenced by national monetary considerations that may prevail at any given time'. ASA does not intend to do business in the USA, and does not expect to be subject to US income or capital gains tax. The investment policy is governed by certain restrictions, including an obligation to invest over 50 per cent in value of its assets in ordinary shares of South African gold mining companies. At the 31st May 1980, 79 per cent of the portfolio value was in gold shares.

There are now several investment trusts, unit trusts and mutual funds specialising in gold or gold related investments. Some are restricted to holding only shares, others are allowed to operate bullion accounts and all have different percentages of exposure to the gold market. It is not possible to provide a definitive list of these here except to say that several are listed in the *Financial Times*, along with the addresses of the fund managers. Many others, which any professional advisor should be able to nominate, also exist. The advantage of these investments is that someone else manages the portfolio for a small charge and they enable the small or regular investor to gain some exposure to the sector where transaction charges (brokers' commission, stamp duties etc.) would be an unacceptably high percentage of the total investment stake. These charges, however, are not altogether avoided by the fund; they are merely diluted by economies of scale. The value of the fund is also a function of the managers' investment skill.

Marketability of gold related investments must also be borne in mind. It is usually not physically possible to sell half a 400 ounce gold bullion bar but it is possible to sell a few Krugerrand coins. Gold futures contracts can be sold speedily, shares are usually readily marketable but this varies with individual stocks, whilst trusts may take longer to dispose of. This means that the investor should recognise the marketability aspect for both flexibility in overall strategy and as a precaution in case of forced realisations.

South African Gold Shares

The alternatives open to investors wishing to hold a stake in the South African gold mining industry include:
  — gold mining companies
  — mining holding companies
  — mining houses
Before evaluating the relative merits of investment in these various types of companies, it is necessary to have a close look at the underlying assets of each category.

Gold Mining Companies — the assets of such companies comprise a lease to mine gold from a particular area and the necessary equipment and infrastructure to exploit the gold contained therein. The most significant feature of these assets is the finite life – gold mines do not normally retain a portion of their annual income for the purpose of purchasing and exploiting new lease areas, nor do they charge depreciation. The ultimate asset value of a gold mine is therefore zero, although there will normally be a repayment of capital at the end of a mine's life. The break-up value of a gold mine can be significant, with about 3 per cent of the total gold produced over a mine's life being recovered on break-up.

Mining Holding Companies — the assets of mining holding companies typically comprise a portfolio of shares (with a varying gold mine content) and, in some cases, also mining leases. The mining holding company overcomes the potential problem arising from the wasting assets of the gold mines by retaining income for investment in developing mines. The gold content of the mining holding companies' portfolios varies, reaching 100 per cent in the case of Anglo American Gold Investment Company Limited (Amgold).

Mining Houses — the assets of such companies comprise their investments in mining and industrial ventures, both domestically and abroad. The financial strength, depth of technical expertise, and the large holdings of mining leases are other important assets which are not normally associated with the other two types of company. The wide spread of interests and the regenerative nature of a mining house's business tends to result in the appreciation of its asset base. The underlying assets of such companies are therefore quite different from the wasting assets of the gold mining

companies. The contributions made by gold to the total value of investments of the mining house varies considerably – from about 20 per cent in
the case of Johannesburg Consolidated Investment Company Limited to
about 80 per cent for Gold Fields of South Africa Limited.

On account of the finite life of gold mines it is necessary to make provision for the redemption of investments made therein. In fact, when comparing investment in gold mines with that in mining houses it is also
necessary to make provision for the capital appreciation foregone as a
result of not investing in the (ideally) infinitely lifed mining houses.
Historical data indicates that the capital appreciation which can be anticipated from a mining house in the long term will range from 5–10 per cent
p.a., and that it will probably be closer to the top end of that range. Such
capital appreciation obviously has a significant impact on any investment
decision involving a choice between the two types of gold investment –
mining companies versus mining houses.

Table 6.4 shows the appropriations which must be made from the total
return of gold mine investments in order to account for their finite life.
Different assumptions as to life of mine and the rate of capital appreciation of the alternate mining house investment have been made as follows:

Table 6.4. Rate of capital appreciation.

	0% p.a.	5% p.a.	10% p.a.
Life of mine 20 years	1.0% p.a.	2.6% p.a.	6.6% p.a.
25 years	0.5% p.a.	1.6% p.a.	5.0% p.a.
30 years	0.2% p.a.	1.0% p.a.	4.0% p.a.

Note: A discount rate of 15% p.a. has been used in the formulation of the above table.

It is evident from Table 6.4 that the impact which the finite life of a mine
has on total investment return can be considerable. this fact must, however, be considered in conjunction with the knowledge that it is mining
ventures (mainly gold) which generate the bulk of the mining houses
income.

The role of the mining house can be more than adequately assumed by
the prudent investor who reinvests a fair proportion of his income. He will
benefit from the relatively high dividend yields on gold shares (see Table
6.5) without being burdened by the heavy overhead structure of a mining
house. Furthermore such an investor is not automatically tied to all the
past investments of the mining house, some of which may have a poor or
mediocre outlook. The tendency for mining houses to retain major investments, despite circumstances which would cause knowledgeable smaller
investors to liquidate their holdings, is a major inefficiency which cannot
easily be overcome.

Risks involved in Gold Share Investments

There is no doubt that the risk involved in investment in a mining house is considerably less than that involved in the direct investment in a particular gold mining share. Risk, moreover, can be considerably reduced by the inclusion of a number of gold mining shares in a portfolio, particularly if the major constituents of the portfolio are also producers of uranium. The relative risks involved in the two types of gold investment are now given under the categories of mining risk and market risk.

Mining Risk — In any mining operation there is always an element of risk concerning the ability of the mine to meet planned output. Failure to do so can arise from either of two types of situations. The first type relates to an ongoing failure to meet planned recovery levels due to lower than expected *in-situ* grades, lower than expected percentage recovery, etc. Risk of this nature is normally minimised by the very conservative assumptions on which the mining houses base their feasibility studies on prospective mining projects.

The second type of mining risk relates to the possibility of fires, explosions, floods, pressure bursts etc. Despite tremendous advances in mining technology such disasters will continue to occur. Fortunately such setbacks very rarely affect more than a small portion of the mine. Good mining practice calls for a mine to develop an excess of ore reserves over immediate requirements. This is done to cater for both setbacks of the nature described above and fluctuations in the gold price. It is, therefore, highly unusual for the above occurrences to have a major impact upon earnings. Furthermore, such occurrences are covered by insurance.

The risk relating to mining operations can be diversified away to a large degree by incorporating a number of quality gold mines in the portfolio. It is in fact likely that the mining risk of a carefully constructed gold portfolio could be lower than that of the gold mining portfolios of some mining houses.

Market Risk — relates to the performance of the gold bullion and, where applicable, uranium markets. The gold mining shares are obviously more closely geared to the gold price than are the mining houses. The latter's exposure (with regard to the value of investments) varies from about 20–80 per cent at current prices and financial gearing through debt is rarely significant. In gaining closer gearing to gold price movements, the gold shares obviously forego their ability to rely on other sources of income during periods of a relatively low gold price. The earnings of gold shares, therefore, fluctuate more markedly than do those of the mining houses.

Table 6.5. Dividend payment record and current yields (June 1980).

Mines	Price in U.S.$	When Declared: Interim	Final	Last Four Dividends (½ yearly) (expressed in U.S. cents)			(most recent payment)	Current** yield %
Blyvooruitzicht	14½	Dec.	June	58	46*	77	22*	13.7
Bracken	3⅛	March	Sept.	28	28*	33	36*	22.1
Buffelsfontein	35	Dec.	June	126	92*	142	192*	9.5
Doorfontein	12⅛	Dec.	June	34	23*	47	24*	5.8
Durban Deep	24	June	Dec.	—	57	47*	140	7.8
East Driefontein	22½	June	Dec.	46*	86	65*	146	9.4
E.R.P.M.	24	June	Dec.	—	11	12*	110	5.1
Elandsrand	7⅜	July	Jan.	—	—	—	—	—
Elsburg	4⅝	June	Dec.	6*	9	9*	28	8.0
Ergo	7¼	Oct.	April	—	29	30*	104	18.5
Free State Geduld	47½	April	Oct.	213	218*	282	584*	18.2
Free State Saaiplaas	6⅛	April	Oct.	—	—	—	—	—
Grootvlei	8⅜	June	Dec.	18*	25	43*	68	13.2
Harmony	18½	Sept.	March	43*	63	101*	203	16.4
Hartebeestfontein	63	Dec.	June	201	126*	342	366*	11.2
Kinross	9⅞	March	Sept.	37	37*	50	103*	15.5
Kloof	27½	Dec.	June	29	34*	94	98*	7.0
Leslie	2¾	March	Sept.	16	16*	21	36*	20.7
Libanon	26½	Dec.	June	69	58*	118	61*	6.7
Marievale	4	June	Dec.	66*	44	47*	55	25.5
President Brand	35	April	Oct.	98	136*	216	394*	17.4
President Steyn	31¼	April	Oct.	57	77*	141	344*	15.5
Randfontein	61	June	Dec.	230*	287	295*	427	11.8
St. Helena	31	March	Sept.	126	146*	208	375*	18.8
S.A. Lands	5⅝	July	Jan.	—	29	24*	24	8.5
Southvaal	19	—	Jan.	—	66	—	171	9.0
Stilfontein	16⅛	June	Dec.	18*	58	41*	122	9.5

Table 6.5. *(continuation)*.

Mines	Price in U.S.$	When Declared: Interim	When Declared: Final	Last Four Dividends (½ yearly) (expressed in U.S. cents)			(most recent payment)	Current** yield %
Vaal Reefs	54	July	Jan.	115*	207	224*	390	11.3
Venterspost	11½	Dec.	June	23	17*	35	79*	9.9
Welkom	11	April	Oct.	46	50*	81	160*	21.9
West Driefontein	74	Dec.	June	287	230*	490	366*	11.6
Western Areas	6¾	June	Dec.	9*	14	14*	43	8.4
Western Deep Levels	36¾	July	Jan.	75*	95	112*	275	10.5
Western Holdings	58½	April	Oct.	259	319*	450	830*	21.8
Winkelhaak	24	March	Sept.	87	94*	149	239*	16.1
Zandpan	10⅜	Dec.	June	34	21*	58	61*	11.4
Gold Investment Companies								
"Amgold"	78	Aug.	March	115*	173	207*	434	8.2
GFSA	68	Jan.	Aug.	98	81*	188	159*	5.1

* indicates interim payments
** Current yield based on last two dividends
Source: *Grieveson Grant Mining Quarterly*, June 1980.

Factors affecting the Profitability of Gold Mines

There are several factors which determine the profitability of a gold mine and may be listed as:

Gold price received
Exchange rates
Production or milling rate
Grade of ore mined/recovered
Working costs
Capital expenditure
Sources of other income
Taxation and Lease payments.

The first two items are non-controllable and have been discussed both in this chapter and Chapter 5. In this Section the other factors are examined.

Milling rate

The tonnage milled refers to the tonnage of ore processed which is normally less than the tonnage mined due to waste sorting at various stages. Milling plant capacity is usually known but production bottlenecks can arise at the hoisting or stoping (mining) stages for a number of reasons. Generally, South African mines have an excellent record of achieving production targets. However, as production is shifted between different parts of the mine or different reefs, temporary fluctuations can and do occur.

Although the reef itself may contain high gold values, by the time the gold has been extracted the average gold value will have been severely reduced. This is because it is not just the reef that is mined. The companies must for two reasons take more than the reef itself: firstly, the gold reef is not necessarily constant in direction and following it exactly would present insuperable technological problems and secondly, the miners would not be able to work unless there was adequate headroom. As it is, working with headroom of only four feet and often less is most uncomfortable. A certain amount of gold, although not much, is also lost in the reduction plant due to the inefficiencies in grinding and processing.

Recovery grade

The calculation of future recovery grades is probably the most important and at the same time most difficult of all the factors that affect a gold mine. It is the most important because a small change in the grade can have a large effect upon profits. It is the most difficult because one can only obtain indications of the amounts of gold contained in rock thousands of feet under the surface.

The method used to calculate future grades is by using all the information available to the best advantage. Before any mining can be effected, various tunnels (different forms of which are known as cross-cuts, drives, raises and winzes) must be opened up. This is known as development and is used to prepare the ground for future mining. As a large proportion of development is on the reef, the grade of the material extracted to create the tunnels can be calculated. In the current workings, the tunnels are close together and an accurate estimation of the grades between the tunnels can then be made. Further away, of course, the estimates become less accurate as there are fewer tunnels. At the same time, the trend of the grades can give one an idea of what to expect. In addition, diamond drill holes from either surface or from the existing workings can give an indication of what grades the overall area may contain.

As mentioned in Chapter 5 all South African mines are bound by law to mine ore to the average value of their ore reserves, i.e. they are forbidden to mine only rich ore and leave poorer but economically viable ore behind. This ensures that all the gold is extracted that can be, but it also facilitates the estimation of recovery grade for evaluation. On the other hand, with the rise in the gold price, many sections of ore are now economic whereas they had not been in the past. The result has been that the mines have been able, and in some cases obliged, to lower their grades in order to extract more ore with a lower gold content. This regulation results in the grade falling when the gold price rises which naturally lowers profit margins. The amount of this grade reduction also depends on the method of working employed. Mines utilising longwall mining methods, e.g. Kloof and Western Deep Levels, have far less flexibility as to where the bulk of ore will be mined and this means that, in the short term at least, the grade is more or less 'fixed'. This is an important consideration when the gold price moves quickly.

Working costs

Working costs include all costs charged to the mining and milling of gold ore and are usually expressed as Rand/ton milled or $/oz produced. The latter is usually a more reliable investment guide for comparative purposes as it excludes the exchange rate and grade variables. Table 6.6 illustrates how gold mining costs have moved. The largest element in costs is labour (50 per cent) followed by power (20 per cent) and stores (20 per cent). In the thirty years prior to 1970 the average working costs per ton rose at rates between 3 per cent and 5 per cent per annum. In the mid-1970s costs rose dramatically, rising from R13.8 per ton milled in 1974 to R23.87 in 1977, which sharply reduced the profitability of gold mining at that time. The rate of increase in working costs on gold mines was held below the rate of inflation in South Africa throughout 1979. Working costs per ton milled increased from R27.14 in 1978 to R30.18 in 1979. The average cost in mid-1980 was R34.00/ton.

The main reason for the big rise in working costs in 1974/75 was a surge in the black mineworkers' wages. The minimum wage for surface workers in 1973 was 55 cents a shift, or roughly R14.30 a month. It shot

Table 6.6. How gold mining costs per ounce expressed in
dollar terms have moved.

	Quarter Ended December 1976	Quarter Ended June 1980	Annual % Increase June 1979/June 1980
Blyvoorultzicht	124,81	172,06	37,9
Bracken	114,47	227,60	98,8
Buffelsfontein	158,79	220,54	38.9
Deelkraal	—	445,44	—
Doornfontein	147,73	189,32	28,2
Durban Deep	242,15	346,80	43,2
East Driefontein	38,71	89,13	63,7
E.R.P.M.	267,96	349,02	30,3
Elandsrand	246,14	273,11	11,0
Free State Geduld	94,03	138,71	47,5
Free State Saaiplaas	256,09	409,88	60,1
Grootviel	140,15	241,47	72,3
Harmony*	204,12	291,63	42,9
Hartebeestfontein	122,20	174,05	42,4
Kinross	122,97	167,48	36,2
Kloof	81,60	120,16	47,3
Leslie	168,45	301,02	78,7
Libanon	139,18	186,09	33,7
Loraine	310,47	484,67	56,1
Marievale	133,63	278,43	108,4
Pres. Brand	112,96	142,56	26,2
Pres. Steyn	135,90	180,63	32,9
Randfontein	163,31	277,66	70,0
St. Helena	98,60	140,89	42,9
South Vaal	112,61	154,49	37,2
Stilfontein	170,33	197,11	15,7
Unisel	—	241,53	—
Vaal Reefs	138,84	187,50	35,0
Venterspost	259,56	321,75	24,0
Vlakfontein	141,00	188,17	33,5
Welkom	169,90	263,78	55,3
West Driefontein	59,76	88,26	47,7
Western Areas	181,86	313,11	72,2
Western Deep Levels	89,93	130,38	45,0
Western Holdings	87,96	134,65	53,1
Winkelhaak	84,41	124,92	48,0

Source: *Grieveson Grant Mining Quarterly*, June 1980.

up by 65 per cent in the next two years. The rate of increase then slackened, moving ahead little faster than the inflation rate, but has been accel-
erating again with 1980 labour costs approximately 25 per cent above
1979 levels. The wage gap between black and white workers has narrowed
appreciably, from 18:1 in the early 1970s to 7:1 in 1980.

Following the labour shortage of 1974, the Chamber of Mines tried to reduce the dependence from South Africa's neighbouring countries. As can be seen from Table 6.7, in 1974 only 25 per cent of the labour came from within South Africa. In 1978, it was in excess of 50 per cent. The major sources of migrant labour are now Lesotho and Botswana.

Table 6.7. Where the workers come from.

	1974	1975	1976	1977	1978
Angola	2,904	3,220	2,356	645	306
Botswana	18,008	22,819	24,244	24,676	21,009
Lesotho	72,343	83,306	86,781	92,875	97,462
Malawi	93,992	8,517	182	6,131	19,799
Mozambique	87,626	100,103	56,404	41,667	33,874
Rhodesia	6	4,332	20,166	15,910	13,049
South Africa	91,793	129,090	173,581	224,622	224,660
South West Africa	941	1,308	2,550	2,757	1,500
Swaziland	5,631	8,313	9,817	9,696	9,299
Total	373,244	361,008	376,081	418,979	420,958

Source: *Financial Mail.*

Mechanisation in the gold mines has been slowed down both for technical reasons and because the cost and quality of available labour rendered mechanisation unprofitable. The peculiar technical problems found in South African gold mines, hard abrasive rock and deep, narrow dipping and faulted seams, mean that machines developed for conditions elsewhere are inappropriate and require adaptation. There are two potential non-technical impediments to extensive mechanisation in the mining of gold. The more mechanised the mining methods, the fewer unskilled and the more semi-skilled and skilled workers (of which the country is desperately short) are required. This creates the danger that new jobs will be closed to Africans by the powerful white miners' Union if it fears that the jobs or safety of members are threatened. However, in the new open cast coal mines nearly all drivers of earth-moving equipment are black and the simpler parts of artisan jobs have been Africanised. The 'blasting certificate' is monopolised by whites, but the more mechanical coal mining methods involve less blasting. The other impediment is the Government policy of limiting the permanent settlement of African miners and their families near the mines, the official maximum being 3 per cent of African employment.

However, South African mining technology in coping with deep, high temperature, hard-rock conditions has made it a world leader in the areas of underground ventilation and mechanical raise boring.

Capital Expenditure

Initial capital expenditure is the capital spent in bringing a new mine to production (including the working costs). Thereafter the major items of capital expenditure are new shafts, housing and extensions to the reduction plant. The replacement of equipment worn out during mining is also capital expenditure and although the amounts are usually comparatively small these costs will continue to be incurred throughout the life of the mine and can be considered as ongoing capital costs. As with labour costs, capital costs have also been increasing. The cost of building a second Elandsrand is now more than R450 million, against the R258 million spent on bringing the mine to full production over the last decade. The capital costs of establishing a new mine can be divided proportionately as: shaft sinking 30.6 per cent, accommodation 21.1 per cent, treatment plant 14.0 per cent, equipment 12.8 per cent, sundries 9.1 per cent, services 7.4 per cent and refrigeration 4.9 per cent.

Because no depreciation is charged against profits any expansion or development of the mine to new areas increases capital expenditure. This can largely be offset against the mine's tax liability. In 1979 capital expenditure on operating South African gold mines totalled R689 million or 10 per cent of total operating profit.

Sources of other income

The major item here is the by-product, uranium. The amounts of uranium contained in the material that is brought to the surface may be very small and could not be mined economically for its own value. However, as the costs of mining and crushing have already been paid for by the gold mining, the extraction of uranium becomes viable. Other income can include such items as sulphuric acid (e.g. West Driefontein) and income from shareholdings (e.g. Vaal Reefs, Western Holdings and Free State Geduld).

Taxation and lease payments

Payments to the State by gold mining companies are of two similar, but independent, parts. The lease payment recognises the fact that all precious minerals are vested in the South African government and is effectively a royalty for extraction rights. The income tax payment takes the form of a conventional corporation tax. Both payments are levied on profits and have the special feature that the rate of payment varies according to profitability year by year. This is effected through the lease formula and the tax formula. The formulae vary according to the classification of the mine. A new mine does not pay tax until all initial capital outlays have been reccovered from the profits. In 1979 the average tax rate payable on the South African gold mines was 48 per cent but after the recent rise in the bullion price several mines have been paying up to 65 per cent. Details of tax and lease formulae are given in the *Mining Journal Quarterly Review of Gold Shares.*

North American Gold Shares

Canadian Gold Mines

Canada is the free world's second largest producer, with an output of 49.3 tonnes in 1980, but well behind South Africa's 675.0 tonnes. Total production showed another decline in 1980, down from 51.0 tonnes in 1979, as grades in lode mines were lowered in response to the far higher gold price, further accentuated by the depreciation of the Canadian currency. In 1979, by-production from base metal mines provided a third of the total. Production from lode mines is expected to continue to drift, but this should be compensated for by the establishment of a new series of mines, mainly in Quebec, working low grade disseminated orebodies. By province, Ontario provided 39 per cent of production in 1979, Quebec 29 per cent, British Columbia 16 per cent and North West Territories 11 per cent.

The industry is dominated by Dome Mines which, with its affiliated Campbell Red Lake (57 per cent owned) and Sigma Mines (63 per cent), produce around 350,000 ounces p.a. or 20 per cent of Canada's total production of 1.7 million ounces per year. Dome's own mine in Timmins, Ontario, was established in 1910 and in 1979 produced 95,000 ounces, with operating costs just under $200/oz. The mine is being expanded by 50 per cent and a new deep level shaft is being sunk at an estimated total cost of Can$50 million. The deposits are all of the fracture filling quartz vein type and all are located in the Canadian Shield. Dome's other main asset, which is estimated to provide nearly half of its consolidated earnings in 1980, is its 26 per cent holding in Dome Petroleum. Other major gold mines include Giant Yellowknife, Agnico Eagle, Pamour Porcupine and Cominco's wholly-owned Con mine in the North West Territories. Table 6.8 gives details of Canadian gold mines.

Table 6.8. Canadian gold mines.

	Annual Production of Gold 000's oz	Current Operating Costs in $/oz (1980)
Agnico Eagle	65	115
Campbell Red Lake	180	65
Dome Mines	95	220
Giant Yellowknife	75	220
Pamour Porcupine	135	300
Sigma	60	150

United States Gold Mines

The USE followed Canada as the third largest producer of gold in the free world with an output of 28.3 tonnes in 1979, or nearly half the amount produced ten years ago. In 1980 the USA fell to fourth place. Only two major mines remain in operation: Homestake in South Dakota with 7.6 tonnes of gold produced in 1979 and Newmont Mining Corporation's Carlin in Nevada with 4.1 tonnes. The majority of the balance is derived as a by-product from copper mining in Utah, Arizona and Nevada. Some reversal in the fall in output is expected in the next few years, as low grade, disseminated zones especially in Nevada are brought to production, starting with Freeport's 6.2 tonnes p.a. Jerrit Canyon mine from mid-1982.

Homestake is the continent's largest producer of gold with an output of 246,000 ounces in 1979. Proven reserves alone represent a mine life of twelve years and an accelerated programme to delineate additional reserves and explore new areas has been initiated. As mentioned in Chapter 5 Homestake has recently made a new gold find in California. The average production cost in 1979 was $250/oz. The company's other main gold interest is a 48 per cent holding in the Mount Charlotte mine in Western Australia which produces around 110,000 ounces p.a. at a cost of $125/oz. Homestake's other interests include silver, lead, zinc and uranium.

Newmont's Carlin open pit produced 134,000 ounces of gold in 1979, at an operating cost of around $150/oz. Proven reserves, including the new Maggie Creek orebody some 14 miles away, are adequate for a minimum of ten years working, but the potential for more reserves in the area is considered to be good. Newmont also owns 70 per cent of the Telfer open pit gold mine in Australia which produces around 175,000 ounces p.a. from high grade ore at over 1/3 ounce per ton. In 1980, gold operations are estimated to provide nearly a quarter of Newmont's net earnings, with domestic and international copper mining activities around 50 per cent. Other company activities include oil, gas and a 27.5 per cent holding in Peabody Coal, the nation's largest producer of steam coal.

Australian Gold Mines

Since the discoveries of 1841, Australia has produced more than 195 million ounces of gold. In the decade 1851–1860, 40 per cent of the world's production came from Australia and even in the 1890–1910 period Australia accounted for 20 per cent of world production. With the Western Australian discoveries of the 1890s, Australian production built up to a peak of 3.8 million ounces in 1903. This was also Western Australia's peak year with production being 2.3 million ounces. To date, over 75 per cent of Australian production has come from Victoria and Western Australia. Of the Western Australian production of 70 million ounces, approximately 50 per cent has come from the Golden Mile lodes at Kalgoorlie. In contrast with Western Australia, over 60 per cent of Victorian production has come from alluvials.

In 1979, Australia produced only 595,000 ounces of gold in total, largely from major operations at Telfer (BHP — Newmont), Tennant Creek (Peko Wallsend), Mount Charlotte (Homestake (US), Gold Mines of

Kalgoorlie and Poseidon) and Norseman (Central Norseman Gold); this is equivalent to less than 2 per cent of the world output.

At present, only three sizeable mine developments in Australia have made firm decisions to go ahead – Fimiston expansion (Poseidon/GMK), North Kalgoorlie and Hill 50's venture. It must be remembered that gold is not uncommon in Australia but generally the deposits are very small and gold in commercial quantities is inevitably scarce. Thus, many companies are producing interesting bore-hole intersections at present but very few prospects are likely to reach the production stage. There is no tax on gold mining profits in Australia at present but this may change.

Evaluation, Assessment and Performance Measurement

To keep in touch with the gold bullion and share markets an investor needs to keep a close watch on the financial press. To obtain specialist information investors usually contact a stockbroking firm which has a research department covering gold mining shares or subscribe to some of the specialist gold newsletters which are available. Probably one of the most widely respected publications is the *Mining Journal Quarterly Review of South African Gold Shares.* This analyses all the South African gold mines on a quarterly basis, comments on the gold market, produces earnings estimates for the major gold mines and provides historical financial and production data for each mine. This is complemented by the *International Gold Mining Newsletter* which covers up-to-the-minute events in the gold mining industry round the world and is airmailed monthly by the same company. The Chamber of Mines of South Africa produces a quarterly analysis of gold mining results.

In assessing the investment merits of any gold share it is important to understand dividend yields (*see Appendix I*). For Australian and North American gold shares yield is less important and the price/earnings ratio (*also Appendix I*) is the most common criteria. Occasionally one will see Net Present Value of gold shares quoted. Simply, these are the future earnings of the mine discounted back to today's money values up until theoretical exhaustion of ore reserves. With a volatile gold price and South African gold share yield (to the non-South African investor) of around 20 per cent Net Present Value is of little relevance. For mining houses the net asset value (usually with investments at stockmarket valuations) indicates how cheaply/expensively the investor is purchasing the underlying assets.

In the short term investors frequently like to know how a gold share price has behaved relative to the gold price or other gold shares by looking at charts. These will not tell an investor what is going to happen in the future but they do show how the shares have behaved in the past. If a gold share price has not been performing as well as others it is as well to remember that it is not consequently 'cheap'. There will be a good reason why it has been under-performing, i.e. there has been an underground fire or something else has happened to offset its income producing ability. It is as well to find these facts out before investing. A rough idea of the volatility of a gold share price is given by comparing the share price high and low over a period with the gold price range over the same period. This will provide the investor with a rough guide to a share's volatility but watch out for currencies and, again, past performance is not necessarily any guide to future performance.

Assessing your performance or success at investment depends very much on your objectives; after all, a profit is a profit. You may wish to compare all your investments against the gold price or against some measure of inflation. The most common yardsticks are the gold mines indices. The *Financial Times* produces a daily gold mines index based on sterling prices and the Johannesburg Stock Exchange produces a rand gold mines index, as does the *Rand Daily Mail*. Unfortunately, all these suffer from currency distortions when compared against a gold price determined in dollars. For Canadian gold shares the Toronto Stock Exchange produces the T.S.E. gold index.

The total return on holding a gold share comprises two separate elements – the income or dividend stream and the remaining capital value of the shares. In addition, since part of a mine's earnings are retained in the company for shareholders, i.e. not all net profits attributable to shareholders are distributed as dividends, the value of the shares should, in theory, grow, but in mining shares this also depends on the amount of the return of capital element associated with a finite ore reserve life. It is normal to include dividends received as part of the total return and value the shares at regular intervals. Hence a gold share portfolio should be constantly growing, whereas the buying and selling of bullion or gold coins depending more on the investor's trading skills has no directly associated income stream.

These then are the major economic factors needed to take into account when evaluating gold shares. The technical risks of investing in gold mining can be seen to be numerous. They include changes in ore grade, the imperfect knowledge of ore reserves which means uncertainty regarding mine life, and natural disasters such as floodings and subsidence. In addition, to these risks, there are currency risks and the political risk of investing in a country which has seen such major domestic disturbances as Sharpeville and Soweto.

The political future of South Africa remains uncertain. The Government remains dominated by Afrikaaner nationalists. However, within the country as a whole, only 60 per cent of South Africa's 4.5 million whites are Afrikaaner against 40 per cent for 'English speakers'. Non-whites, whose political rights are rigidly controlled, number 23 million. This disparity between the white ruling élite, and the black and coloured population will become greater in the next twenty years. Ninety per cent of school children currently are black or coloured.

Gold Futures Markets

An increasingly important form of gold investment particularly for US investors has been gold futures markets. A description of the markets themselves can be found in Chapter 7. In this Section the principles underlying investing in gold futures are examined.

A futures contract for gold is a legally binding instrument to buy or sell a designated quantity of gold at a specific time period in the future, at a price agreed upon today. The contract details the standards the gold must meet in order to be acceptable for delivery. An important aspect of a futures contract is the possibility of leverage and this is discussed further in Chapter 7.

Futures prices for gold bullion normally trade at a premium. In other words, if the 'spot price' or 'cash price' (the price quoted for bullion for immediate delivery) is $530, the price for six month gold may be $580. This premium takes into consideration two factors: current interest rates and the outlook for the gold price. The interest rate factor is a constant one. If the futures market did not take it into account, the cash price and the price for one-year delivery would be the same. Industrial and private holders of gold would simply sell their gold into the cash market for $530 per ounce, and invest their cash float at the going interest rate for twelve months. This simultaneous supply to the cash market and demand in the twelve months market would immediately bring that premium back into force. The second factor, that of the price outlook for gold, is obviously not a constant one. It varies considerably and makes premiums on future delivery dates for gold fluctuate. When the gold price is under heavy downward pressure, premiums contract; when gold is on a steady upward curve, they expand.

There are two types of futures markets participants; speculators and hedgers. Speculators invest in the market hoping to correctly anticipate and take advantage of price swings. Hedging, on the other hand, involves use of the futures market to protect the producer, processor or handler of a commodity against adverse price movements that might affect his merchandising profit. The following two examples will help to illustrate how differing views as to the future of the gold price can be made use of by hedgers. Assume a gold mining company expects gold prices to decline in nine months time (from, say, the current $580 per ounce). They would be well advised to place a short hedge in the futures market. Assume that it is September at the moment, and that the company makes its heaviest sales in July. The company feels that $560 an ounce would assure it of a reasonable profit after covering all costs. Accordingly, a sale of June futures is undertaken at that price. Now assume it is June, and prices have in fact declined. The company would then sell its newly mined gold on the cash market at $550 an ounce, and simultaneously buys back its June futures contract at $551 an ounce. The transaction known as a short hedge would look something like this:

A Short Hedge.

Cash Market		Futures Market	
September 1			
Anticipated sale price based on June future	$560/oz	Sold June gold @	$560/oz
June 1			
Sold gold @	$550/oz	Bought (offset) gold @	$551/oz
Profit from futures	= $ 9/oz		
Realised price	= $559/oz		

Thus the company has made a profit of $9 per ounce in the futures market and has sold its inventory at $559 per ounce, nine dollars more than it would have received had it not hedged. Had prices increased rather than declined during the time of the operator's hedge, his loss on his futures transaction would have been offset by his gain in cash sales. Of course, in the case of rising prices, without a hedge his cash sale would have resulted in a larger profit. This is an option to speculators. In this case, the company realised that by hedging it gave up the opportunity for the extra profit, but at the same time it was insured against obtaining a dramatically lower price.

To take an opposite example, assume that a jeweller needs gold six months ahead and fears that the gold price is likely to rise. In this case he can undertake what is known as a long hedge. Rather than purchase the actual inventory of gold in September the jeweller could forward price his gold by buying March futures contracts. On September 1 he orders March delivery of gold at $550. On March 1, he buys his needs on the cash market at $565 and sells back his futures at $564. The $14 an ounce profit in the futures transaction more than offsets the $10 increase he paid for the actual gold. The transaction known as a long hedge would look something like this:

A Long Hedge

Cash Market		Futures Market	
September 1			
Ordered gold for early		Bought March gold @	$550/oz
March delivery at projected			
price of	$550/oz		
March 1			
Bought gold @	$565/oz	Sold (offset) gold @	$564/oz
Profit from futures =	$ 14/oz		
Realised price =	$551/oz		

Therefore, entering into a contract for a stated date and later closing it out is a simple and secure method of:

(a) hedging: using a futures contract as a surrogate for a deal later to be executed in another market and taking a profit or loss on the futures deal to offset an extra cost ('loss') or an extra gain ('profit') on the deal in the other market – a risk avoidance operation.

(b) speculating: (analysing and taking a risk which can subsequently be controlled) on a future rise or fall in the price – a risk taking operation.

With a futures contract one can take advantage of a gold price increase while simultaneously earning interest on your money (this is in contrast to holding bullion). Thus by shrewd judgment (combined with the leverage effect) one can benefit from either a gold price rise (by buying gold futures) or a price fall (by selling futures). On the negative side, one is forced to deal in certain contract sizes (*see Table 6.2*), one can only deal for delivery on a certain contract date and one can only deal during set trading hours. The problem of dealing within set trading hours is being reduced by the worldwide trend towards opening gold futures markets. Existing future markets currently exist in New York, Chicago, Sydney, Singapore, and were joined by Hong Kong in August 1980. Discussion about a London gold futures market, to be sponsored by the bullion houses and members of the London Metal Exchange is currently taking place.

Gold Call Options

By purchasing a futures contract an investor is promising to purchase gold at a specified price and specified date in the future. By purchasing an option, an investor does *not* promise to purchase the gold but he does have the *right* to purchase gold at a specified price at a certain date in the future. When an investor purchases an option he pays a premium to the seller of the option. This premium buys him a contract that states that for a certain period of time he has complete freedom of choice to demand delivery of gold at the agreed price, referred to in the market as the striking price. Therefore he has purchased a contract which, though limited in time, enables him to enter the gold market with only a small commitment of funds and with the complete confidence that the maximum loss he can expect is the cost of the premium paid. There are no margin requirements for the buyer of an option; he simply pays the premium.

This premium is affected by the laws of supply and demand. If at any one time there are more purchasers in the market than sellers, the premium will have a tendency to increase. It is, therefore, evident that the premium is influenced by the general tendency of the market. Indeed, when the market in general is 'bullish', it is most likely that one will find many buyers around, whereas sellers of options will tend to withdraw, thus pushing the premium higher. Conversely, when the bullion market is 'bearish', the number of investors likely to sell options to protect themselves against a decrease in the value of their investment will tend to increase and this will tend to lower the premium. The premium is also influenced by other factors, such as:

The distance between the market price and the striking price of the option. The higher the market price of bullion is relative to the striking price, the higher the premium will tend to be.

The time left to maturity
Clearly the longer the time left to the date of maturity, the higher the premium will be, while at the date of maturity, the premium will become exactly equal to the spread between the market price of bullion and the striking price, if the spread is positive. The premium will be zero if the spread is negative.

The Winnipeg Commodity Exchange was the first in the world to open when it instituted trading in call options on its 100 ounce gold futures contracts on April 30, 1979. These options are traded at striking prices in multiples of $20 and are cleared through the clearing house of exchange. A market in traded gold options is expected to open in Amsterdam in the near future. In the USA, a market for call options is made by Mocatta Metals Corporation of New York who feature a wide range of striking prices and maturity dates. Options are for 100 ounce units. Mocatta Metals Corporation is also active in call options for South African Krugerrands. Trading is in units of 100 pieces. Europe's largest dealer in options on gold bullion is Valeurs White Weld S.A. of Geneva, a wholly owned subsidiary of Credit Suisse First Boston. Valeurs White Weld options are now also available in the USA through appointed Future Commission Merchants (at the moment they are First Boston Corporation, New York, Bache Halsey Stuart Incorporated and International Trading

Group San Francisco). A slight disadvantage to North American investors was that the Swiss options were denominated in five-kilogram units (equivalent to 160.75 troy ounces) but 100 ounce units have recently been introduced. Striking price and premiums are quoted on a US dollar per ounce basis to facilitate trading. Valeurs White Weld, in an informative free brochure, have outlined the benefits derivable from various strategies open to purchasers and sellers of call options, the relevant parts of which are reproduced here.

Benefits from alternative strategies open to purchasers of call options

(a) Leverage effect with limited risk

The buyer expects a double result by buying an option when he believes that the underlying bullion will increase in value during the time of the option. On the one hand he hopes to earn the full capital gain while investing a very small fraction of the value of the bullion, thereby benefitting from the leverage effect and on the other hand, he knows from the start that if his 'bullish' forecast were to be wrong, he would stand to lose no more than the small amount of funds needed for the purchase of the option, whereas he could have lost a far greater sum if he had owned the gold bullion itself or bought it for future delivery on a futures market. An example will make this clear.

Example
On May 16th 1980, when the spot price of gold was $515 an ounce, an investor buys a November 1980 options for 100 US metal account gold at $610 an ounce for which he pays a premium of $38 an ounce. This option will therefore cost him $3,800.

On the 15th September gold bullion is priced at $666. The option which is still valid for over two months has increased in value to $70 per ounce. The investor is, therefore, in a position to resell his option for the price of $7,000. In this example he would have made an 84 per cent gain and limited his total risk to $3,800 whereas had he bought the bullion he would have had to invest $51,500 which would have been at risk and he would have made but a 29 per cent gain on it.

Onviously, the investor in this example may also elect not to resell his option but to exercise the option and take physical delivery of the gold bullion, but there would be no interest in doing so for as long as the option were selling for more than the difference between the market price of gold and the striking price.

Another way for the investor of protecting his current profit while maintaining his option position open until maturity would be to sell an option for the same maturity but at a higher striking price against the option he holds. Such a sale is exempted from margin requirements. Again an example is useful.

Example

On September 15th 1980, with gold trading at $666, the investor of the above example has four courses of action:

1. He may sell back his option for $70/oz. thereby realising a net $32 profit representing a flat 84 per cent gain on his investment.

2. He may hold on to his November 610 call until maturity in the hope that the gold price will rise even further. He runs the risk, however, that if gold stays at $666 until the end of November, his net profit will then be only $19 or 47 per cent flat on his investment (Spot price $666 minus striking price $610 minus original cost of option $38).

3. He may lock in his profit by holding on to his November 610 call and selling the underlying gold on the futures market for the date of maturity. Since there remain 2 months until maturity, he would possibly obtain $682/oz. on his forward sale and be assured of a minimum net profit of $34. (Forward price $682 minus striking price of $610 minus $38 for original cost of option.)

4. He may hold on to his original option and sell against that one a November 285 call option for which he might receive a premium of $29/oz. He then in effect will hold a 'spread' which he can hold until maturity for a net cost of $9/oz. (Premium paid $38 minus premium received $29.)

As long as the gold price remains above $619 he will make a $1/oz. profit for each dollar above $619 up to a maximum profit of $71 if gold stays or rises above $690. His total risk is $9/oz. should gold fall to or below $610.

(b) Buying an option and investing the difference

This strategy provides a certain insurance against loss of the total amount of the premium. The investor who was willing to invest in bullion outright may invest instead in the option and invest the difference in a short-term security or deposit. He will therefore earn an interest on the major part of his investment, which can be offset against the cost of the option.

(c) Buying an option in anticipation of a cash flow

Options may also be used to secure the price of the metal which the investor wishes to purchase in the future in anticipation of a cash flow at that time. For example, an investor may be expecting a sum of cash from the maturity of bonds or Certificates of Deposit or from the sale of property. He may also have decided that he wishes to place some of these funds in gold. In expectation of a rising gold market in the intervening period, he may wish to establish the price he will pay for the bullion today by the purchase of an option.

(d) Hedging a short sale in the futures market

An option may be a good hedge to a short sale. For the investor who has sold bullion short on a future commodity exchange, the purchase of an option provides adequate protection against the possible increase in the price of bullion. It establishes the maximum price he will have to pay for the bullion in order to satisfy his obligation on the short sale.

(e)   Options as a way of hedging inventory requirements
Industrial users of gold, uncertain of the future evolution of the price of the metal and unwilling to take the risk of entering into a forward purchase contract, will find the option a convenient way to set a maximum price on their future requirements while benefitting from any decrease in price.

Benefits from alternative strategies
open to sellers of call options

(a)   The sale of options as a source of return on investment.
The long-term holder of bullion receives neither dividend nor interest on his investment. Therefore it is clear that in periods of stable or moderately rising prices the bullion holder may find an advantage in selling options on his investment. The proceeds or premium from the writing of such options may at times provide the investor with a return well in excess of that yielded by time deposits of equivalent maturity.

(b)   The sale of options as a protection against a decline in the market value of bullion.
Again an example makes this clear.
Example
Let us assume the case of an investor holding 1200 ounces of gold at the average price of $660 per ounce while the actual market price is at $666 per ounce and is not expected to go higher in the near term. Our investor fears that if the market should break it would find its next support level between $620 and $630. To protect his investment he may wish to sell 12 options against his holding. Let us assume he can obtain $50 per ounce for a six-month option at the striking price of $710. He would then incur no loss on his position until the market price of bullion reached $610 per ounce. ($660 average purchase price minus $50.) On the other hand, his bullion would only be called if the bid price in the market rose above $71 per ounce, in which case he would then have made at least a profit of $100 per ounce. ($710 minus $660 plus $50.)

(c)   Alternative strategies based on the sale of options.
If it is possible to achieve a certain amount of downside protection by writing a call option on one's bullion, can one not increase such downward protection by writing more than one call option to achieve greater price protection? Indeed, that strategy, known as 'variable hedging', involves writing one option covered by the investor's long position in bullion and one or more options that are uncovered. The additional downside protection provided by the additional premium income must be weighed, how-

ever, against the risk of incurring a loss on the option or options that are uncovered.

Thus call options provide leverage, clearly defined risk, provide a direct investment in gold (unlike investing in gold shares) and involve no storage costs. Their disadvantage is that dealing can only be done in certain contract sizes, trading hours and maturity dates. In addition, they tend to be less liquid and slightly more expensive than futures contracts.

Conclusion

In circumstances of instability in the economy, the attraction of gold as an investment is greatly increased. Gold has a unique combination of qualities. It is a real asset in the same way as other commodities, land or buildings are real assets. It does not depend for its value on credit or on estimates of its future earning power. Gold is at the same time, a liquid asset. It is possible in almost all circumstances to change gold into money, or into other kinds of money, at an hours notice and in extreme circumstances, gold can be used to purchase other goods when paper money is not acceptable. A gold bar purchased this afternoon, which has been produced by an acceptable refiner, can be negotiated for cash tomorrow morning in Frankfurt, Zurich, London, Hong Kong or New York. Gold is therefore the only investment which is almost 100 per cent real and almost 100 per cent liquid. During a period of inflation real assets with low liquidity such as property can prove to be dangerous investments; it may be impossible to sell them while inflation is causing violent fluctuations in interest rates and money values. Government bonds which have high liquidity but do not represent real values, are also a dangerous investment in a period of inflation. Of the three functions ascribed to money – use as a unit of account, as a medium of exchange and as store of value – it is the last which gold has increasingly performed in recent years. Although clearly affected by the choice of dates, the average price in dollars of gold during 1980 was some 1,470 per cent higher than its average in 1970. Few assets can compete with this return on investment.

APPENDIX I
INVESTMENT CRITERIA TO BE USED IN EVALUATING GOLD SHARES

There are two forms of yield associated with ordinary gold shares. The first and most commonly used is the dividend yield, and when investors or journalists speak simply of a share's yield, this is the one they mean. The second is the earnings yield.

DIVIDEND YIELD
This expresses in percentage terms the income return the investor expects

to get from investing in a share. This expectation will be based on the amount of dividend paid by the company on its shares, and yields are normally calculated on the basis of the last full year's dividend. The yield of a share is calculated from two statistics: the amount of dividend paid per share, and the market price of the share. Dividend yields normally refer to the income produced by an investment before tax has been deducted. The gross dividend yield is derived from the following formula:

$$\frac{\text{Gross Dividend x 100}}{\text{Current Share Price}} = \text{Gross Dividend Yield (\%)}$$

EARNINGS YIELD

The earnings yield involves the same form of calculation as the dividend yield, but refers not to the dividend actually paid but to the profit earned by the company that is attributable to ordinary shareholders. This is the residual profit in the company's profit and loss account after all prior claims on its profit have been met; these include taxation, minority interests and preference dividends. The significance of the earnings figure is that it shows the amount of dividend the company could pay to its shareholders if it chose to distribute all of its full year's profit. The earnings expressed either as so much per share or as an earnings yield, provide the basis for calculating one of the most important investment ratios — dividend cover.

Dividend Cover

The dividend yield of an ordinary share is always a matter of expectation rather than fact. Companies pay dividends from their profits, and for the most part they aim to achieve a steady increase in profits over the years from which regular dividend increases can be paid. In practice, however, progress is usually more fitful, with poor years intermixed with the good, and dividend records may reflect this. If the company has been in the habit of paying out a high proportion of its profits each year as a dividend, then a fall in profits is likely to cause the dividend to be reduced. If, on the other hand, the company has followed a more conservative dividend policy, paying out to shareholders significantly less than the full amount of its year's earnings and using the balance to develop its business, then the probability is that the year-to-year effect of trading fluctuations can be absorbed without the need for a reduction in dividend.

Dividend cover is a measure of the conservatism of a company's dividend policy, and therefore of shareholder's income security. It is measured in either of the following ways:

$$\frac{\text{Earnings per share}}{\text{Dividend per share}} \quad \text{or} \quad \frac{\text{Earnings yield (\%)}}{\text{Dividend yield (\%)}}$$

Cover of 1 means that the company is paying out the full amount of its earnings as dividend, and therefore has little or no room to absorb a fall in earnings without reducing the dividend, while cover of 2 or more represents a strong measure of dividend security.

Price/Earnings Ratio

The price/earnings ratio, usually abbreviated to P/E ratio or P.E.R., was adopted in the mid-1960s as an indicator of the value of shares comparable with the dividend yield. It is calculated in exactly the same way as the name suggests:

$$\frac{\text{Current share price}}{\text{Earnings per share}} = \text{P/E ratio}$$

The information used is the same as for the calculation of the earnings yield, which is arrived at by:

$$\frac{\text{Earnings per share x 100}}{\text{Current share price}} = \text{Earnings yield (\%)}$$

and it will be found that the two ratios, the P/E ratio and the earnings yield are reciprocals; multiply them together and the answer will be 100. This relationship is potentially important and valuable, for if the P/E ratio is known the earnings yield can be found quite simply by the formula:

$$\frac{100}{\text{P/E Ratio}} = \text{Earnings yield}$$

and vice versa. On this basis, it is possible to calculate the level of dividend cover where only the yield and P/E ratio are published.

As mentioned above, the P/E ratio is to a great extent the counterpart of the dividend yield. Whereas the yield focusses attention both on the present dividend and on what investors are expecting future dividends to be (a low yield points to expected growth, while a very high one reflects fears of a reduction), the P/E ratio highlights a company's earnings and investors' expectations for those earnings.

The P/E ratio is often said to measure the number of years' earnings of a company that can be purchased at the current price. If the price of a share is 200 cents and earnings 25 cents, then the investor is said to be paying for eight years' earnings at this price. This, however, ignores the fact that the company's earnings are 25 cents for one year only, and it is clearly impossible to forecast what the total of its earnings will be over the next eight years. The P/E ratio is nevertheless a very useful shorthand method for investors to indicate the expected earnings prospects for a company relative to those of other companies.

If the average P/E ratio for shares is, for example, 12 then a company whose growth prospects are above average might be expected to have a P/E ratio of above 12: perhaps 13 or 14 if its prospects are only moderately better than the majority, or as much as 20 or 25 if they are substantially better. In effect this means that investors are prepared to pay a higher price for this share than for most others because they believe it represents a more attractive investment. The opposite will be true for companies expected to produce a worse earnings performance than the majority, and P/E ratios in these cases will stand at an appropriate level below 12.

The importance of earnings growth is that it makes possible a higher rate of dividend growth. It is normal therefore for a share that has a relatively high P/E ratio to have a relatively low dividend yield, reflecting expectations of above-average future dividend growth; and for a share with a relatively low P/E ratio, reflecting poor earnings prospects, to have a relatively high yield.

APPENDIX II
GOLD INDEXED BONDS

The French government has, over time, issued two gold linked bonds. The 'Pinay' bond, so called after the French Finance Minister who introduced it has a coupon of 4.5 per cent on its par value of FFr 100. The redemption price is linked directly to the rise in the price of the FFr 20 Gold Napoleon, the favourite French way of investing in gold. The Pinay can be bought on margin (20 per cent down) in the French market, it is purchasable in small amounts, and it is reasonably certain that the Government will honour the redemption terms because the bond is being continually redeemed via a sinking fund and has very widespread ownership.

The 7 per cent Emprunt is a French State fifteen year loan. It is widely known as the 'Giscard' for the same reason as the naming of the 'Pinay'. It is redeemable on 16th January 1988, either at par or according to certain guarantee arrangements; interest is paid yearly on 16th January, subject to deduction of French withholding tax — 25 per cent at November 1980 — but there is 10 per cent tax or no tax if there are exemption agreements with the country of the holder and France. There is no tax on capital gains for foreign holders. There are two guarantee arrangements, the first based on the Common Market Unit of Agricultural Account (U.A.A.) and the second on gold. The two guarantee arrangements are:

(1) If the French Franc is officially devalued, the coupon and the repayment price are revalued by as much relative to the Franc as the U.A.A.

(2) If any one of five events takes place, this arrangement is replaced by a revaluation by the extent of the rise in the Paris price of the 1 kilogram gold ingot from the FFr 10,483 at which it stood when the loan was issued. The five events are: (i) the U.A.A. no longer applies to the Franc; (ii) the U.A.A. is no longer defined as equal to a quantity of gold; (iii) the value of the U.A.A. has been suspended for a whole calendar year; (iv) the

French Franc is no longer defined as equal to a quantity of gold; and (v) the French Franc has floated for a whole calendar year.

In 1977, the Franc floated for a full year, and the guarantee produced a coupon of F 168.80 (per F 1,000) in January 1978. In March 1978, the ratification of the new Articles of the International Monetary Fund ended the definition of the Franc in terms of gold: the guarantee produced a coupon of F 193.86 in January 1979, and a coupon of F 392.96 in January 1980. It has been estimated that interest and repayments may now cost the French Treasury 100 billion Francs before redemption in 1988 against the original 6½ billion Francs raised on the bonds in 1973.

Chapter 7 discusses the issue in February 1981 of a gold linked Euro-bond.

Chapter Seven

THE WORLD'S GOLD MARKETS

Introduction

The major gold markets around the world tend to coincide with financial centres where a concentration of funds is seeking opportunities for investment. Thus gold trading has traditionally taken place in London, Zurich, Hong Kong and, more recently, New York. Yet the presumption that gold must be traded in this or that centre is growing steadily more obsolete. In the course of the past decade the gold market has become an increasingly global affair. The major trading houses are now equipped to follow the market around the world on a twenty-four hour basis. As the earth turns, the price of gold follows the sun since as the London market closes, the New York markets are then open followed by the Hong Kong markets, and by the time the Hong Kong market closes the London market has re-opened.

The increasing prominence of gold as an investment for troubled times has also encouraged trading to develop in places like Tokyo, Singapore and Frankfurt where wealth is very much in evidence but where there was, until recently, no tradition of gold trading. In addition to dealing through the normal gold markets there has been considerable speculation that the OPEC countries, who are short of gold and anxious to diversify out of the dollar, have been indulging in direct oil for gold barter deals with the world's major gold producers.

This chapter examines bullion trading in the major centres of London, New York, Chicago, Zurich and Hong Kong. Grey and blackmarket gold trading is not discussed. Table 7.1 provides a comparison of the gold contract specifications in these major centres. The gold futures markets were discussed in Chapter 6 and are further discussed in this chapter.

The London Gold Market

London's pre-eminence as a gold trading centre stems from the last century when South Africa's gold started being marketed in London. Although South African gold mines are privately owned, the law stated that all newly mined gold must be sold to the South African Reserve Bank as

Table 7.1. Comparison of gold contract specifications.

	The London Gold Market	The Zurich Gold Pool	Hong Kong Gold and Silver Exchange Society
Trading unit	400 ounces troy, fine	400 ounces troy, fine	50 taels gross = 60.165 ounces
Value dates traded	Spot and any forward date	Spot and any forward date	Spot
Delivery units	Bars of 350–430 ounces troy, fine weight .995 minimum fineness	Bars of 350–430 ounces troy, fine weight .995 minimum fineness	Bars of 5 taels gross .990 minimum fineness
Place of delivery	London vaults,, or elsewhere by arrangement	Zurich vaults, or elsewhere by arrangement	Specified banks in Hong Kong or at Exchange
Daily limits on price movement	None	None	None
Minimum price fluctuation	None	None	None
Trading hours local time	Mon to Fri 0900–1700 with Fixings at 1030 and 1500 (Johnson Matthey dealers normally cover approx. 0715–1915)	Mon to Fri 0930–1200 and 1400–1600	Mon to Sat 0930–1230 and 1430–1600
Trading hours London time		0830–1100 and 1300–1500	0130–0439 and 0630–0800

Trading on other markets is given in relation to GMT. When London is on BST, add one hour to the London time for all other market hours, allowing also for any local clock.
Source: Johnson Matthey Bankers Limited.

Table 7.1. *(continued).*

Winnipeg Commodity Exchange		New York Commodity Exchange (Comex)	International Monetary Market of the Chicago Mercantile Exchange (IMM)
'Standard' 400 ounces troy, fine	'Centum' 100 ounces troy, fine	100 ounces troy, fine	100 ounces troy, fine
January April July October up to 15 months forward	February May August November up to 15 months forward	Spot and following two months. February, April, June, August, October, December up to 17 months forward	Spot, and March, June September, December up to 18 months forward
Gold Certificates issued by specified banks in Toronto in amounts of 400 and 100 ounces troy fine weight, representing bars .995 minimum fineness		Bars of 100 ounces troy fine ±5% 995 minimum fineness Specified depositories in New York City	Not more than 3 bars against each 100 oz contract, minimum 31 ounces troy fine weight, .995 minimum fineness Specified depositories in New York City and Chicago, Illinois
U.S. $10.00 per ounce troy (no limit on final day for each month)		U.S. $10.00 per ounce troy (no limit during contract month)	U.S. $10.00 per ounce troy (no limit during contract month)
5 U.S. cents per ounce troy		10 U.S. cents per ounce troy	10 U.S. cents per ounce troy
Mon to Fri 0815—1330		Mon to Fri 0925—1430	Mon to Fri 0825—1330
1415—1930		1425—1930	1425—1930

agent for South Africa's Treasury. The Reserve Bank sold the gold in London through the Bank of England.

The stage for the first fixing of the gold price was set in 1914 when Britain abandoned the gold standard. All through the war the Bank of England had been buying South Africa's entire gold output at £4. 4s. 11d. per fine ounce. But in 1919 the pound sterling was devalued against the US dollar. Anxious to benefit from the higher price available in the USA, South African mining interests persuaded the Bank of England to allow N. M. Rothschild to market their gold, about half of the world's output, 'at the best price obtainable'.

Prior to 1917, daily gold sales were rather informal. A representative of N. M. Rothschild and Sons trekked through the 'city' each day offering gold for sale. The firm acted both as agent for the Bank of England and for itself. But as volume increased it seemed easier and more dignified to invite the other bullion dealers to Rothschild's office for a formal offering ceremony. It was in this way that the gold 'fixings' started. Thus on September 12, 1919, the date of the first 'fixing', London emerged as a place where gold was priced as distinct from a place where gold was refined, assayed, dealt in, shipped to or stored. The first fixing, somewhat curiously, took place by telephone. The price was set at £4. 18s. 9d. per fine ounce, a premium which was roughly in line with the degree to which the dollar had been revalued against the pound. Shortly afterwards it was agreed that the four dealers, Mocatta and Goldsmid, Pixley and Abel, Samuel Montagu and Rothschilds should meet regularly at Rothschild's offices.

Mocatta and Goldsmid's history stretches back to the late seventeenth century when it was silver broker to the Bank of England and when London was taking over from Amsterdam as the world's leading centre for gold trading. Mocatta and Goldsmid is now a subsidiary of Standard Chartered Bank. N. M. Rothschild was founded in 1804 and assembled the cash needed to finance the closing phases of the Napoleonic war. Pixley and Abel was set up in 1852 and Samuel Montagu in 1853; their emergence coincided with the Californian and Australian gold rushes, much of whose gold was handled in London. Pixley and Abel merged in 1957 with the older firm of Sharps and Wilkins, which was independently present at the first fixing. The result was Sharps, Pixley. Shortly after the first gold fixing Johnson Matthey became a regular member.

The essence of the fixing is that one price is found which satisfies a large number of sellers in one many-sided transaction. The advantage to the customer is that the spread is narrow, as befits what is essentially a matching exercise. The seller gets the fixing price. The buyer pays the fixing price plus one quarter of a per cent. A much larger order for gold can be accommodated during the fixing than during normal trading. At 10.30 am and 3.00 pm each business day the chairman of the fixing, from Rothschilds, opens the ritual with the phrase 'Gentlemen, we will start at such and such a price'. This opening price is reported back to the dealing rooms of the five firms. They alert potential customers worldwide, who, taken together, provide each firm with a net requirement to buy or sell a number of Good Delivery Bars of about 12.5 kilogram of gold each.

The price the Rothschilds' chairman suggests reflects pre-fix trading activity. Each of the participants has a small Union Jack on his desk. If all the bullion brokers are sellers, or all buyers, or if their orders do not match the price is adjusted accordingly and the whole process is gone through again. Sellers specify the quantity offered; buyers do not indicate the size of bids. When all representatives are satisfied they lower their flags and the price is fixed. A dealer needing more time can halt the proceedings by simultaneously saying 'flag up' and raising his Union Jack. When the fixing figure is final, it is transmitted throughout the world by electronic news services, by telephones and telex, and, finally, by being printed in the financial newspapers. Although no dealer is bound by the London fixing, the figure serves as an accurate reflection of supply and demand in the world market and is an important indicator to all market participants in Europe and abroad. Only a small percentage of worldwide daily gold turn-over is transacted at the two fixings. Nevertheless it is extremely important, since many private contracts specify a gold price based on a future date's fixing. Occasionally the match of buying and selling orders is not perfect and settlements must be arranged *pro rata*.

The London market functions in accordance with several set trading guidelines. Dealers' specifications of what bars are acceptable, what fineness they need to have, and where delivery can be made are used. Most leading gold dealers maintain a 'bullion account' with one of the five London dealers through which they can settle transactions between themselves. It is quite conceivable that a Far East gold trader and a bank in Canada would agree on London delivery when trading gold with each other. Since 1968, the London market has quoted gold in US dollars per ounce. Table 7.2 gives the specifications for a good delivery bar. Most gold traded in London is in 400 ounce bars. This size bar came into existence because it was the heaviest bar South African labourers could carry. From this origin the bars became the standard size used by the Bank of England and the precious metals industry. Most central bank monetary stocks are also in 400 ounce bar form.

A gold bar physically shows its origin and history. Its shape indicates whether it was cast in the USA or elsewhere. US bars are rectangular bricks, 7 inches long, $3\frac{5}{8}$ inches wide and between $1\frac{5}{8}$ and $1\frac{3}{4}$ inches thick. Most bars cast outside the USA are trapezoidal. Another type of bar occasionally seen in the USA are the so called 'Hershey bars'. These are created when an amount of gold too small to make a full bar is left in the smelters crucible at the end of the casting process. Since the purity of the gold varies between different pourings, this leftover metal cannot be added to other pourings and must be cast in a separate bar.

Table 7.2. Specification for a good delivery bar.

The unit of dealing is a bar conforming to the following specification:

Weight	Minimum gold content: 350 fine ounces
	Maximum gold content: 430 fine ounces
	The weight of each bar shall be expressed in ounces troy in multiples of .025 of an ounce and must turn the scale at the weight indicated.
Fineness	Minimum 995 parts per 1,000 fine gold
Marks	Serial Number
	Stamp of acceptable Melter and Assayer

Each bar, if not marked with the fineness and stamp of an acceptable assayer, must be accompanied by a certificate issued by an acceptable assayer stating the serial number of the bar and the fineness.

If a bar bears more than one assay stamp, preference shall be given to the British assay. Where a bar bears no British assay the lowest figure will be taken.

Gold said to be 1,000 fine shall be marked down to 999.9 fine.

Bars must be of good appearance, free from surface cavities or other irregularities, layering and excessive shrinkage, and must be easy to handle and convenient to stack.

A buyer cannot stipulate any particular brand of bar. If a bar included in the specification is tendered and it does not suit the buyer's requirements, the cost of melting and/or refining will be at his charge.

Bars not conforming to the specification may be sold on the Market, but the seller will be charged with the cost of making them good delivery.

The simple spot quotation for gold the world over is for gold 'loco-London', that is, for delivery in London. This eliminates complications of freight cost and insurance from the price equation; it also means *de facto* that much of the clearing of transactions in physical gold takes place in London. In addition the members of the London Gold Market control the Good Delivery list. This is the list of melters and assayers whose marks are acceptable as assurance of the weight and quality of gold bars. The world still accepts the London Gold Market as arbiter in this quality control.

There was no real competitor to London as a gold market until 1968. As described in Chapter 3 the market was suppressed between 1925 and 1931 by a second British attempt to impose a gold standard on the United Kingdom. The market was shut down by the Second World War and was reopened in 1954 with the price standing at £12. 8s. an ounce. The Bretton Woods agreement had established gold as the ultimate basis of reserve currencies and the IMF had banned the use of gold for non-monetary investment purposes. But inexorably the pressure on the gold price mounted and this led to the reopening of the London market, as other European centres were already trading gold at premium prices. The London market re-established supremacy because it was in London that South African gold output, private demand and the intervention of the Bank of England were matched to hold the price at around the official $ 35 per ounce.

For fourteen years the central banks maintained this price against gradually increasing speculative pressure. In 1961, the London Gold Pool was formed, in which the central banks pooled their efforts to hold the line, and this reinforced London's pre-eminence. But by 1968, with sterling already devalued and the dollar suspect, the pressure from speculators had become no longer sustainable. In March the Gold Pool was abandoned and, under the Washington Agreement, a two-tier market in gold was introduced with an official price for central bank transactions and a free market price for everybody else.

The London market declined in importance as a result of the suspension of the London gold pool. When gold's official price was abandoned, the market was ordered to remain closed for two weeks. During the same two weeks, the Zurich market remained open and quickly became used to the fluctuations in price which now prevailed every day. During that period the South Africans transferred all their gold sales to Zurich through the newly estabilshed Zurich gold pool. The big three Swiss banks, Credit Suisse, Swiss Bank Corporation and Union Bank of Switzerland, effectively replaced the central banks as buyers of last resort, at market prices, during the hiatus in London. Britain's policy on arms bans to South Africa helped Zurich and it was not until 1970, and a Conservative Government, that a proportion of the South African output returned to London. A London gold futures market is planned to open in the near future.

The Zurich Gold Market

Switzerland has long been an important centre for gold trading. Its role as a world market used to be limited to gold coins but today it is also active in the bullion market. As mentioned earlier, in the late 1960s Zurich picked up much of London's business. South Africa began channeling the majority of its transactions through the newly established Zurich gold pool which combined with the resources of Credit Suisse, Swiss Bank Corporation, and Union Bank of Switzerland. In 1972, when the USSR brought sizeable amounts of gold to the market, Wozchod Handelsbank in Zurich looked after the major portion of sales. Business has shifted back and forth between Zurich and London over the past few years, but both centres are highly respected for their professionalism and expertise.

Bullion is traded on a market basis with the same regulations for good delivery as were illustrated in Table 7.2. Coins fall into two categories. The first consists of 'numismatic coins', ie. those minted before 1810, which are traded at individual prices not directly related to the metal price. The second comprises current coins issued since 1810; these are traded on a market basis, like bullion. According to their premiums over gold, current coins are further divided into three groups: semi-numismatic coins, which have high premiums; genuine current coins, with normal premiums; new mintings, with low premiums.

The Swiss have a lot of advantages over other centres in the gold market. They have the advantage that they have a number of private accounts of investors for whom they can buy and sell, and many of the investors who they service are people who are concerned about keeping the purchasing power of their assets and who are interested in the anonymity that gold ownership provides — anonymity in the sense that one does not have to register as a shareholder, one does not have Securities Exchange Commission problems, one does not have to vote and decide what to do in a proxy fight, decide whether you should tender or not tender when somebody makes a rights offering etc. In this sense for the kinds of investors serviced by the Swiss banks gold provides convenient investment. In addition to that, of course, the Swiss banks have relatively cheap sources of money available; sometimes even within their own banks, they are able to get funds at quite low rates. Their central bank until a few years ago permitted them to hold gold as part of their reserves *vis-à-vis* the central bank, and this was another very great benefit that the Swiss banks had in the gold market, which is now becoming less important. In addition they are substantial shareholders of Swissair, which is one of the large transporters of gold around the world.

In Switzerland all purchases of gold coins and bullion by Swiss nationals or non-residents are subject to Swiss turnover tax (5.6 per cent upon purchase) when physically acquired in Switzerland as of January 1, 1980. The tax is not applicable to items held in collective custody acquired prior to January 1, 1980 even when delivery is asked for after that date. However, exempt from Swiss turnover tax, are gold bullion (minimum quantity 1 kilogram = 32 oz) and several gold coin purchases through precious metal accounts (claim accounts). Dealing is usually conducted on a spot basis though forward transactions are also quite normal.

The forward price is based on the spot buying or selling rate prevailing at the time the deal is concluded, increased by interest calculated for the lifetime of the contract at current rates in the Euromoney market. Forward contracts are not subject to the retail sales tax except in such cases where the holder of a long contracts asks for physical delivery of the metal involved against payment. On conclusion of a forward deal the customer is requested to put up a margin in cash or in any other readily negotiable assets acceptable to the bank undertaking the transaction and to sign the special form 'Charge'. For the time being, the margin amounts to 50 per cent of the equivalent of the contract, interest included, whereby the bank reserves the right to increase this percentage and ask for additional cover at any time. If a margin call is not met within the stipulated time, the bank furthermore reserves the right to liquidate the contract in the market and to sell the assets pledged and offset the proceeds against the loss incurred on the contract. As a rule, a forward contract may be liquidated at any time prior to its expiration date, either by delivering or withdrawing the metal involved against payment at the contracted rate or by liquidation of the contract at the market, whereby the resulting profit or loss (i.e. the difference between the contracted rate and the market rate) is accounted for with value liquidation date. The unused interest is added to or deducted from the liquidation rate depending on whether the liquidated contract is a long or short one.

In order to simplify gold transactions and reduce the movement of heavy gold bullion to a minimum especially when large turnover is involved gold business is often transacted via a 'claim account'. Instead of safekeeping fees, the customer pays a commission, which in the case of gold amounts to 0.1 per cent per annum of the highest credit balance. The account holder does not buy actual gold bars, but acquires a claim on gold. The bank undertakes to hand over, on request, the amount of gold credited to the account.

In addition to the large 'good delivery' bars the Swiss banks also produce small bars and wafers. Small bars are cast in a mould and are available in various sizes, as can be seen from Table 7.3. In Europe and in many

Table 7.3. Weights and dimensions of small bars.

1,000 grams	(=	32.151 ounces)	117.0 × 52.0 × 9.0 mm
500 grams	(=	16.075 ounces)	86.0 × 38.0 × 8.5 mm
250 grams	(=	8.038 ounces)	58.0 × 30.0 × 8.0 mm
100 grams	(=	3.215 ounces)	45.0 × 25.0 × 5.0 mm
50 grams	(=	1.608 ounces)	34.0 × 19.0 × 4.0 mm
10 ounces	(= 311.035 grams		58 0 × 32.0 × 9.0 mm
10 tolas	(= 116.638 grams)		45.0 × 27.0 × 6.0 mm

overseas markets the one kilogram bar is the most popular. It is sold in all standard finenesses, namely 995/1000, 999/1000 and 999.9/1000. The smaller units, however, are only made in fineness of 999.9/1000.

Wafers, unlike bars, are not cast but stamped and embossed and provided with a high lustre finish. Table 7.4 illustrates the weights and dimensions in which wafers are traded. Wafers are exclusively available in a fineness of 999.9/1000.

Since 1972 the USSR has added to the supply of primary gold to Zurich, selling a large part of its variable annual output through the Zurich-based Wozchod Handelsbank. But in recent years it seems that

Table 7.4. Weights and dimensions of wafers.

100 grams	(=	3.215	ounces)	51.2 × 27.6 × 3.9	mm
50 grams	(=	1.608	ounces)	43.1 × 23.3 × 2.8	mm
20 grams	(=	0.643	ounces)	35.6 × 19.3 × 1.7	mm
10 grams	(=	0.322	ounces)	28.0 × 15.0 × 1.4	mm
5 grams	(=	0.161	ounces)	20.7 × 11.0 × 1.3	mm
5 ounces	(=	155.5175	grams)	54.6 × 33.0 × 4.9	mm
1 ounce	(=	31.1035	grams)	31.5 × 17.0 × 3.2	mm
1 ounce	(=	31.1035	grams)	approx. 28.0 × 2.75	mm
½ ounce	(=	15.5518	grams)	25.0 × 13.5 × 2.6	mm
½ ounce	(=	15.5518	grams)	approx. 23.0 × 2.2	mm

Zurich's position as the focus of the supply of new gold has been some-what diminished; first, because South Africa has diverted a small pro-portion of its annual sale back to London (some observers say about 20 per cent and some more to New York) and second, because there is a suspicion that the USSR is now selling through other outlets, including direct placement with Arab investors. Swiss bankers are reputed, never-theless, to play an important part in selling Soviet gold, even though these sales may no longer be exclusively through the Wozchod Handelsbank.

These suspicions are brought out in Swiss customs statistics which were published in late 1980. These showed that some 145 tonnes of gold were transferred from Switzerland in the first seven months of 1980 by Iraq (45 tonnes) and Kuwait, the United Arab Emirates and Iran (71 tonnes). This is well up on the combined imports of 45 tonnes in 1978. Thus these four countries have increased their imports of newly purchased gold and of previously bought stocks of gold. This is a good example of the war chest motive mentioned in Chapter 5. In contrast, however, Saudi Arabian imports stood at 8 tonnes, well down on the average of 50 tonnes im-ported in the previous three years. Overall the imports of Middle East countries of gold from the Middle East have risen suggesting that these countries prefer to now hold their gold outside Switzerland. Figure 7.1 illustrates the recent trend of Swiss gold imports and exports. In addition, Soviet gold exports to Switzerland, as can be seen from Figure 7.2 have considerably declined, particularly since mid-1978. Between January and August 1980 there were no Soviet gold sales in Zurich. Later in the year there were sales of 34 tonnes in September and October. The Soviets may, however, be disguising these sales, operating via some of their satellite countries.

The United States Gold Market

New York's development as a gold trading centre began only in January 1975 when US citizens were given the right to hold the metal. Gold trading then spent two and a half years in the doldrums before taking off

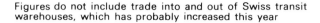

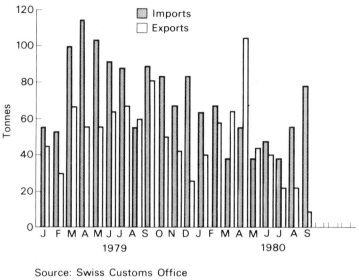

Source: Swiss Customs Office
Financial Times

Figure 7.1. Swiss gold trade.

into the stratosphere. It has concentrated itself in two areas: gold coins, especially Krugerrands, and futures markets. The futures market discussed below is characterised by a leverage effect. Thus with a margin deposit of, say, 500 to 1000 dollars the investor could participate in a gold price increase not solely of that value of gold but of 15 or 20 times as much.

The New York dealer market is large, active and operates similarly to the over-the-counter securities market. It is composed of large international bullion finance houses, precious metal dealers and some brokerage firms. There are four firms which function as gold merchants: J. Aron, Mocatta Metals, Sharps Pixley Incorporated, which is part of the British firm, and Philip Brothers. Perhaps one should also include one New York Bank, the Republican National Bank which is also in the bullion business.

Although any quantity may be traded, most transactions are in multiples of 100 ounces. The common units are 400 ounces and 100 ounces bars of 0.999 purity. Trading is not restricted to exchange hours, but is conducted throughout the day. Delivery may be made on any mutually agreed upon date. Terms are usually payment on receipt. Prices quoted are in line with the latest London fix. Quotations of different dealers are extremely close.

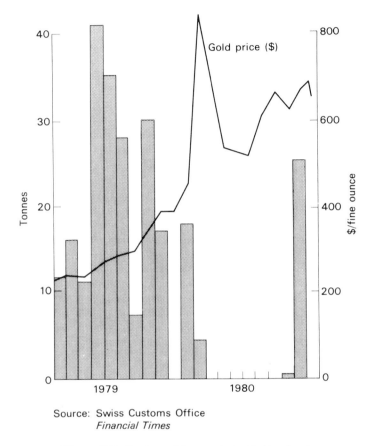

Figure 7.2. Soviet gold exports to Switzerland.

While this market is dominated by gold professionals, a private investor can avail himself of its facilities by telephoning one or more of the participating dealers. A dealer list is in *Metal Statistics*, published annually by the Americal Metal Market.

Gold Futures Markets in the United States

As discussed in Chapter 5 on the last day of 1974 it became lawful again for US citizens to own gold and five exchanges, three in Chicago and two in New York, opened their floors for trading in gold futures contracts. Details of the growth of US futures markets have been taken from *Gold*

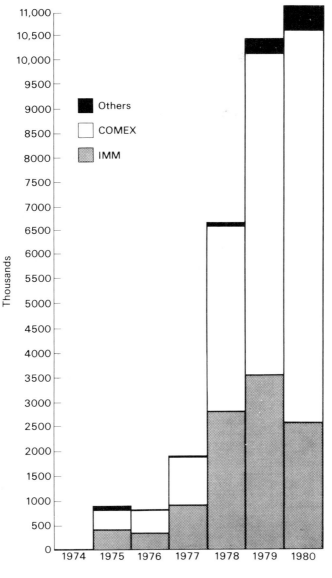

Source: *Gold 1980*. Published by Consolidated Gold
Fields Limited

Figure 7.3. The growth of US gold futures contracts.

1980, published by Consolidated Gold Fields Limited. At the close of the first day 7,475 contracts were traded and over the next twelve months the trading volume was 898,421 contracts. After that start, second year trading in 1976 hovered in the same range with 836,148 transactions, 7 per cent lower than the previous year. With the third year, however, volume began surging upward: 1977 trading 2.3 times ahead of the previous year, with 1,909,789 contracts and 1978 improved on that level by 3.5 times with 6,660,239 gold futures contracts traded. Then, in 1979, market activity soared: heavy demand sent the annual volume of futures traded to 10,412,273 which exceeded all the trades of the previous four years combined, by 107,676 contracts. Thus, as the decade ended, as many gold futures contracts were trading in an average month as were traded in the first full year of trading.

In 1975 Commodity Exchange Incorporated, New York, (COMEX) and the International Monetary Market (IMM) in Chicago handled 89 per cent of the gold futures volume. That pattern has persisted. In 1979, as can be seen from Figure 7.3 with trading volume almost twelve times greater than in 1975, COMEX and the IMM shared 97 per cent of the gold futures business, with COMEX having 63 per cent and the IMM 34 per cent. The remaining 3 per cent of gold volume was accounted for by the Chicago Board of Trade and Mid-America Exchange.

An extremely lucid account of the mechanics and importance of US gold futures markets has been provided by consolidated Gold Fields in *Gold 1980* from which the next section has been reproduced.

'Customer orders can and do originate anywhere in the world, but to be executed they must be transmitted to the floor of the Exchange. By law, no USA futures transaction can occur outside the designated confines or business hours of the Exchange. Every member firm has either telephone or other electronic equipment on the floor of the Exchange to which its futures business is transmitted. Once transmitted to the floor, all customer orders are executed by a broker in the trading area ('pit') for the specific commodity. The transaction is consummated by 'open outcry' in auction style whereby bidders and offerers compete for the best price. Under normal market conditions, trades are executed swiftly, within two to three minutes after the order is entered, and then reported back through the firm's system. Since time is crucial to futures market customers, the Exchanges keep precise account of the time at which any price change occurs. Customer orders are time-stamped at receipt and confirmation. Once the customer order has 'cleared', the Clearing House of the Exchange becomes the opposite party: the seller for every buyer, the buyer for every seller. Customers need not be concerned about who takes the opposite sides of their trades on the floor, since the Clearing House stands for performance of the contract.

Once a trader has taken a position in the market by buying or selling

Table 7.5. USA Gold Futures Contracts.

	COMEX Contract Commodity Exchange, Inc. New York, New York	IMM Contract International Monetary Market, Chicago, Illinois
Trading Hours (New York Time)	9.25 am–2.30 pm	9.25 am–2.30 pm
Contract or Unit	100 troy ounces plus or minus 5% Minimum .995 fineness	100 troy ounces plus or minus 5% Minimum .995 fineness
Fluctuation Minimum Daily	10¢ per troy ounce	10¢ per troy ounce
Fluctuation Maximum Daily	$25 per troy ounce* Maximum range $50	$50 per troy ounce* Maximum range $100
Delivery	Approved depositories in New York	Approved depositories in Chicago and New York
Initial Trading Months**	February, April, June, August, October, December, up to 30 months**	March, June, September, December, up to 21 months**

*Maximum permitted fluctuations are subject to change by the individual exchanges.
**Plus spot month thus allowing the first three months to trade.
Source: *Gold 1980* published by Consolidated Gold Fields.

one or more contracts, he has two options: firstly to maintain this position until the contract matures and then accept or make delivery, or secondly to offset the contract before maturity by assuming a position equal and opposite to the original trade. In other words, if a trader originally bought one gold contract, he could liquidate or offset by selling that contract; if he originally sold a contract, he could offset by buying it back. In both cases, his liquidating trade must be made for the same delivery month and on the same exchange as the original contract. Approximately 97 per cent of all commitments are offset in this manner, rather than by delivery of the gold itself.

The customer's account is kept current by each member firm on a daily basis. When a customer enters the market, he deposits with his firm a required 'margin'. This represents a security deposit towards any adverse price movement in the market. If the market continues to move against the customer position and the margin money is eroded below a minimum set by the Exchanges, the customer is required to deposit additional money with the firm. If the market moves in favour of the customer position, the customer may withdraw the excess money on deposit in his account.

Because security deposits (margins) are relatively small, futures market positions are highly leveraged. As a result, the Exchanges establish daily limits on the amount which the price of a particular

Table 7.6. Comparison of gold futures and gold spot market dealings.

IMM Futures Market	Dealer (Cash Market)
1. Participants are either buyers or sellers at a single, specified price at any given point in time.	1. Participants make two-sided markets, (quoting two prices that indicate a willingness to buy at the lower price and to sell at the higher price).
2. Non-member participants deal through brokers, who represent them on the IMM floor.	2. Participants deal on a principal-to-principal basis.
3. Market participants are usually unknown to one another except where a firm is trading its own account through its own brokers on the IMM trading floor.	3. Participants in each transaction always know who is on the other side of the trade.
4. Trading is conducted in a competitive arena by 'open outcry' of bids, offers, and amounts.	4. Trading is done by telephone or telex with customer orders coming into each dealer's trading room.
5. Participants include those who are in the gold business as well as those who are not, including large and small 'retail' or individual speculators and investors.	5. Trading is dominated by participants in the precious metals business; access for individuals is limited.
6— Daily price and volume statistics are disseminated publicly by the IMM.	6— Actual prices are not publicly available (applies to daily trading; fixes are exceptions) and no data are available on total volume.
7— The Exchange's Clearing House becomes the opposite side to each cleared transaction; therefore, the credit risk for a futures market participant is always the same and there is no need to analyse the credit of other market participants.	7— Each party with whom a dealer does business must be individually examined as a credit risk and credit limits must be set for each. There may be a wide range in the credit capabilities of participants.
8— Margins are required of all participants.	8— Margins are not always required, depending upon a dealer's assessment of each counterpart to a transaction.
9— Settlements are made daily via the Exchange's Clearing House. Gains on position values may be withdrawn and losses are collected.	9— Settlement takes place two days after spot transactions. For forward transactions settlement occurs two days after the maturity date.
10— A small percentage (usually less than 1 per cent) of all contracts traded results in actual delivery.	10—The majority of trades results in delivery.
11—Generally, trading is in contracts for future delivery.	11—The bulk of trading is spot.
12—Long and short positions are easily liquidated.	12—Forward agreements are not easily offset or transferred to other participants.

Table 7.6. *(continuation).*

IMM Futures Market	Dealer (Cash Market)
13—Standardised dates are used for delivery of gold in the spot month and/or the quarterly months of March, June, September, and December.*	13—Forward contracts can be delivered at any date agreed upon between the buyer and seller.
14.—Contract size is 100 troy ounces of .995 fineness.	14—Dealers normally trade 400-ounce bars of .995 fineness, though other weights and qualities are available.
15—Pricing basis is for Chicago or New York delivery. Specifications of IMM gold contracts allow delivery at seller's option in designated warehouses in either Chicago or New York.	15—Pricing basis is normally for London delivery, though delivery charges for other locations may be included by dealers at customer request.
16—A single, round-turn (in and out of the market) commission is charged. It is negotiated between the broker and customer and is relatively small in relation to the value of the contract.	16—No commission is charged, but the dealer's cost of making the transaction is built into the price and often substantially exceeds that of the futures market commission.

*The Board of Governors may add additional contract months from time to time. Spot months are opened at each interval between the 'regular' contract months.
Source: *Chicago Mercantile Exchange.*

contract can move, up or down, on a single trading day. This serves to brake extreme price fluctuations and gives the participants an overnight chance to cool down and review the market. It also gives the firm a chance to call for new or more margin should it be required. The markets also utilize an 'expanded limit' formula which works to expand the daily limits so that, when necessary, the market can more quickly catch up to world spot prices.'

A Summary of the provisions of contracts traded on COMEX and IMM gold futures markets is provided in Table 7.5.

The major characteristics of the futures bullion market (in this case the IMM) which distinguish it from the spot bullion market are summarised in Table 7.6.

The Hong Kong Gold Market

Hong Kong has emerged as the major market in the East largely because it is the centre through which East Asian countries match their surplus supplies and demands and channel their overall surpluses or requirements on to the world market.

Hong Kong already has three established markets and has recently acquired a fourth. The longest established market is the Chinese Gold and

Silver Exchange which trades gold in *taels* (the Chinese measurement of weight) and in Hong Kong dollars. This is essentially a spot market, although positions can be carried forward indefinitely on a day-to-day basis on the payment of a premium. This makes it a *de facto* futures market as well.

The second market is the 'loco London' market. This is the market which reflects the global spread of the world's main bullion dealers. It is a spot market for gold, prices as for delivery in London, but trading eight hours ahead of the London opening.

The third Hong Kong market is the 'local COMEX', an off-shoot of the New York Commodities Exchange gold futures market. This is an out-of-hours market run informally by some of the big commodity trading houses which trade on the New York COMEX but which have operations in Hong Kong.

The newcomer, which has recently opened, is the Hong Kong Commodity Exchange's controversial gold futures market. The Government of Hong Kong is behind the scheme, hoping that it will reinforce Hong Kong's attraction as a financial centre, but there has been strong opposition from the Chinese Gold and Silver Exchange Society (Kam Ngan). The new market is modelled closely on the system used in New York and trades in US dollars to heighten its attractions for the gold trading establishment. In addition *taels* are being replaced by ounces as the basic trading unit. The opening of the new futures market brings Hong Kong into direct competition with Singapore.

Chapter Eight
THE FUTURE OF GOLD

Introduction

Gold continues to play an important role in the world's economy. High inflation, political and economic uncertainty and currency fluctuations induce investment and speculation in gold. During 1979 and 1980 the price of gold moved from below $300 per ounce to over $850 (over $1,000 on US futures markets), and back to $398 by July 1981. The aftermath of the Soviet invasion of Afghanistan, which forced up the price to $850 per ounce has been followed by a collapse to around $400 per ounce by July 1981. This price fall is the product of an easing of political tensions, the sharp world recession and the rising US dollar backed by record US interest rates in the eighteen months since gold's rapid price increase. In order to appreciate future likely developments in the gold markets it is necessary to examine the likely developments on both the demand and supply side of the gold market. On the supply side one must ask what factors will influence the major gold sellers: South Africa, the USSR and official gold sellers such as the IMF and the US Treasury. On the demand side it is necessary to examine all the factors which alter the demand for gold both by official monetary authorities and by private market participants. Consequently one must look at the possibilities of political instability, fears of worldwide inflation, problems of currency inconvertibility, the impact of OPEC diversifying out of the dollar, pressures for gold remonetisation, i.e. the restoration of an official gold price, the importance of the European Monetary System and the role of rising third world debt with the possibility of large scale banking failures.

The Outlook for Future Gold Supplies

The monthly auctions of the IMF came to an end in May 1980; so a source which, for four years, provided the market with almost 200 tonnes a year, has now dried up. An equal quantity of gold was returned to IMF member states. Conflicting opinions prevented any decision at the Hamburg meeting of the Interim Committee at the end of April 1980, so about 3,200 tonnes of gold still in the possession of the IMF will not yet be sold. The US Treasury held its last surprise auction on November 1, 1979, when it sold about 37 tonnes.

Milton Friedman's suggestion in the *Optimum Quantity of Money* that central bankers desist from 'fine tuning' and instead announce and keep to an immutable money supply growth rate has been adhered to in the gold market. IMF auctions, and then US Treasury auctions of gold have effectively acted to increase the supply of gold in an automatic and well publicised fashion. The announced intention of US Treasury gold sales was to support the dollar in the foreign exchange markets, rather than to increase the predictability of gold supply and hence the investment attractions of the yellow metal The cessation of these auctions has considerably reduced available gold supplies and, given the large losses incurred by the US Treasury, has resulted in little likelihood of their resumption.

Statistics published by Consolidated Gold Fields indicate that the free world gold supplies diminished by some 52% in 1980 as compared with 1979. (1704 tonnes in 1979 and 803 tonnes in 1980.) Over the last decade South African gold production, as distinct from the value of output has declined steadily. As can be seen from Table 5.11, in 1970 gold production peaked at 1,000 tonnes. Since 1970 the trend has been downwards. 1980's output was about 675 tonnes. Despite this fall in output the mines are earning more and more from less and less gold. Grades are also falling. Since 1972, when grades averaged 12.7 grammes a tonne, the average has fallen by about 35 per cent to 7.28 grammes a tonne by 1980. One technical reason for this, discussed in Chapters 5 and 6, is that under the law which tightly controls South African gold mining the companies have to mine closest to 'payable ore' – that is, the grade on which a reasonable return can be made. The South African Chamber of Mines has projected that gold output will remain at around 700 tons per year until 1987, before declining steadily to around 350 tons by the year 2000.

Increased gold revenues (reaching R10.37 billion for 1980 as compared to R6 billion in 1979) and more sophisticated marketing of the metal have transformed South Africa's trade position; current account deficits between 1975 and 1977 have been displaced by surpluses between 1978 and 1980, though non-gold trade deficits have widened. Mr Chris Stals senior deputy governor of the South African Reserve Bank announced in June 1981 that South Africa was likely to sell all its anticipated gold production of about 650 tons during 1981. This marks a departure from South African policy in 1980 when it increased its gold reserves by 66 tonnes, which was mainly the result of retaining gold production as well as unwinding previous bullion swaps with commercial banks. This policy was due to the emergence of a current account deficit in the balance of payments in 1981. The annualised figure from the first quarter of 1981 gives a current account deficit of R1 billion. Mr Stals stressed that South Africa would not reduce its reserves of gold as foreign borrowing was not expected to be a problem.

In 1979 the USSR contributed 229 tonnes and in 1978 a record 410 tonnes to the world supply of gold. The exceptionally high volume in 1978 was probably drawn from accumulated stocks and provides no clue to the yearly capacity of Soviet gold production. Estimates of current and forecasts of future Soviet gold sales given the total secrecy shrouding Soviet gold operations, are virtually impossible to make with any certainty. However Consolidated Gold Fields have estimated a further fall in Communist gold sales (dominated by the Soviet Union) to 90 tons in 1980. Some motives for Soviet gold sales were outlined in Chapter 5.

As with South Africa, the Soviet Union has no wish to flood the gold market thereby bringing the gold price down very quickly. Thus both the Soviet Union and South Africa have a vested interest in a high and rising gold price. This rising gold price is particularly important as gold production is an energy intensive business – typically 1½ tons of ore are needed to produce 1 ounce of gold. The energy shortage which causes energy prices to rise faster than the inflation rate illustrates their need for a rising gold price. There is evidence that the supply of gold has been carefully regulated by official sales of gold in the West, and newspaper reports have suggested an unofficial cartel between the two main producers, South Africa and the USSR.

The rise in price during the late 1970s has encouraged a good deal of exploration activity and several new gold mines are preparing to start up. These include the Lupin project in the North West Territories and the low grade open-pit mine at Detour Lake (Ontario) and Queen Charlotte Island. However in view of the long lead times required to establish new mines it seems unlikely that world production will rise significantly in the next few years. The impact of this on the gold price has to be tempered by remembering, as was pointed out in Chapter 5, that newly mined gold is only a small component in the overall supply picture – making up just over 1 per cent of overall world stocks.

The UK stockbrokers Phillips and Drew, in mid 1980 in their monthly review summarised the factors which could bring downward pressure on the gold price (the bear factors) and the factors which could bring upward pressure on the gold price (the bull factors) for the early 1980s. The bear factors are:

(1) The present oil glut and fall in the oil price.

(2) A secular trend towards recession and tighter monetary policy in OECD countries, leading to a sustained fall in world inflation.

(3) The average cost of production of gold in South Africa of US $210 theoretically allows scope for sales down to this level.

(4) As a result of the high price of gold in the 1970s, non-Communist production is likely to increase by about 20 per cent to 1,140 tonnes between now and 1985 (Mining Magazine survey). It takes approximately five years to bring a new gold mine on stream.

(5) The probable USSR current account deficit for the next few years may require gold sales of 200 tonnes per annum against 50 tonnes in 1980. Industrialised countries will also be in substantial deficit in 1981 which may involve official transfers of gold to OPEC.

The bull factors are:

(1) Inflation, which may be endemic to Western-style democracy. The oil price is likely to rise in real terms even if it does so in staccato fashion. The OPEC balance of payments surplus is likely to continue. Therefore, whether the causal explanation of the increase in the gold price is the expectation of higher world inflation, or direct purchases by OPEC residents, the price is likely to continue its upward trend.

(2) The evidence that fabrication demand is picking up at around the current price. Fabrication demand tends to decline sharply in relation to speculative demand when gold is in a strong uptrend with a corresponding turnround in the mix after the price has fallen or when trading is subdued at lower levels. The revival of fabrication demand suggests that the downside is limited.

(3) Official sales from the USA and IMF are unlikely unless the market becomes severely overheated. If the projected increase in non-Communist production is added to Communist sales of 200 tonnes per annum, supply should not exceed 1,350 tonnes by 1985, which only slightly exceeds the average level of net new supply less official sales for the 1970s.

(4) The evidence of a cartel between South Africa and the USSR, South Africa swap arrangements, and the possibility of Soviet trading swaps, which suggest that the main producers will not want to swamp the market, despite, in the case of South African gold, relatively low production costs.

(5) 50 per cent of annual net new supply is produced in South Africa. If this were perceived to be at risk because of industrial or political unrest upward pressure on the gold price could be intense.

The Effect of Political Instability

After the Soviet occupation of Afghanistan gold climbed to a previously unthought of price of $855 an ounce, four times its quotation of 13 months previously. This happened when there was an exceptionally abundant supply. In 1979 non-communist production of about 960 tons was supplemented by sales from official holdings. The auctions of the International Monetary Fund and the US Treasury, sporadic sales by central banks and additional sales from South African reserves amounted to another 634 tons. With the 229 tons sold by the Soviet Union, more than 1,800 tons of gold found its way to the market. At the same time, industrial demand (mainly for price-sensitive jewellery, and in particular for gold articles used in the Middle East as a traditional means of hoarding) receded by 20 to 30 percent. In spite of this drop in demand, the large supply was absorbed at prices averaging twice as much as those of the previous year, because of investment and speculative demand. Thus a political crisis sparked off a major wave of gold buying. However, political crises are not an automatic recipe for a gold price increase. This crisis gold buying can only occur if the people who are physically threatened have both the wealth to buy large quantities of gold and also believe that their holdings of gold are below a satisfactory level. If the parties concerned have adequate gold in suitable locations then a major crisis could occur without necessarily forcing a substantial rise in the gold price. As was illustrated by the slight gold price movements after the abortive attempt to rescue the USA hostages in Iran, those who may have felt threatened by subsequent upheavals in the Middle East had already acquired adequate precautionary stocks of gold.

As was pointed out in *Gold 1981* the subdued reaction of the gold price to events in connection with the Iran/Iraq conflict, problems in Poland and the attempted assassination of President Reagan support the above view. Thus *Gold 1981* suggests that the gold price will rise significantly in response to political problems only if the crisis can be seen to be leading quickly towards an armed confrontation of the super-powers and/ or there is going to be a major disruption of Middle East oil supplies to the importing industrial nations.

Prospects of Rising Worldwide Inflation

Rising inflation has become a worldwide problem. In the 1950s the weighted average rate of inflation in the OECD countries was less than 2 per cent. It then rose to 4.5 per cent in the 1960s, to 8.2 per cent in the first half of the 1970s. With oil prices on the rise the prospects of bringing down inflation rates appear dim. In this regard it is instructive to quote from *Gold 1980*:

'The energy constraint on the world's production capacity will be a permanent feature of economic life and attempts by powerful groups to obtain real increases in their incomes at the expense of others will result in a continuation of high inflation rates. The oil producing countries now appear to have the knowledge and strength to ensure that their prices will be maintained in real terms and there is the additional possibility that oil production may be reduced. From time to time inflation rates will rise to higher levels than the previous peaks and some governments will find popular support in attempts to reduce inflation by raising interest rates, slowing down the rate of increase in money supply and causing a business recession. This policy will have to be reversed when unemployment rises to a level which changes the attitude of the electorate and makes the governments reassess their priorities.'

In the USA economic recession might lead to a temporary checking of price rises. On the other hand there is the real danger that the recession will be more severe than officially expected and that even President Reagan will be compelled to run to an expansive economic policy instead of fighting inflation.

Writing as early as 1919, while attending the Paris Peace Conference, John Maynard Keynes argued that there was no surer means of 'overturning the existing basis of society than to debauch the currency.' The process of inflation, he warned, 'engages all the hidden forces of economic law on the side of destruction, and does it in a manner which not one man in a million is able to diagnose.' Small buyers, big investors and central banks from the Less Developed Countries (LDCs) need a hedge against this inflation and gold is one commodity they have chosen to use as a hedging device.

Increased Risks for International Banks

Under the impact of inflation and the oil shock the tasks of national and international financing have grown. So too have the risks. The massive increase in the volume of international borrowing has not gone primarily into productive investments. In the main, borrowing has served the important purpose of financing the oil imports of the developing nations so as to keep the wheels of their economies turning. Thanks to this financial assistance, the developing countries have managed to weather the perilous conditions created by the oil revolution. However, the real burden of servicing these debts has been made all the heavier by rising interest rates and the necessity for quicker repayment.

The banks have so far proved equal to the challenge of financing LDC debt and have made the necessary adaptations in their operating policies. But the initiative innovative spirit and venturesomeness of the banks are not infinite. Indeed the danger of a financial collapse triggered by the collective default of several LDCs still remains. The nature of this problem is suggested by data compiled by Morgan Guaranty Trust Company, showing that the twelve major borrowers among non-OPEC developing countries (Argentina, Bolivia, Brazil, Chile, Colombia, India, South Korea, Philippines, Taiwan, Thailand, Turkey and Ivory Coast) will incur a collective current account deficit in 1981 of $ 45 billion, against $ 38.5 billion in 1980 and $ 9.6 billion and $ 22.3 billion, respectively, in 1978 and 1979. However, after allowing for scheduled amortisation of long-term external debt, the gross financial requirement of this group of countries alone is estimated at some $ 62.5 billion in 1981, compared with $ 54.9 billion in 1980.

International banks have been forced to accept that foreign lending presents three main kinds of interrelated risks which are not only growing larger but are not entirely in the banks power to lessen or to eliminate. These risks are country risk, the risk of interference by Governments and the liquidity risk. As Ossola has put it writing in the June 1980 issue of the Banca Nazionale del Lavoro Quarterly Review:

Country risk – ie the risk of lending to an organisation in a foreign country in a currency other than the local one – is essentially a political concept. It materialises when a country is unable or unwilling to pay at maturity. It may materialise even when the borrower is a perfectly solvent enterprise if its Government does not allow the transfer of the amount due and paid in at the borrower's central bank (this particular risk is called 'transfer risk') Some country risks have always been evident (Turkey), some have exploded suddenly (South Korea); others were for long in the pipe line, but warnings about them went unheeded (Iran), or they were latent because of economic policies pursued over a long period (Zaire). Risks in Poland, Brazil and the Philippines are appearing on the horizon. A major insolvency would provoke a chain reaction which could shatter the foundations of the international financial system.

A second important risk connected with international lending which came to the fore during the Iranian crisis, is the risk of Government interference in the market mechanism. As discussed in Chapter 3 President Carter blocked the Iranian assets in retaliation for detention of the hostages by the Khomeini regime. These actions create problems going beyond the original participants and compel banks to face difficult decicions.

The third important risk is due to the principle whereby Euro-banks normally borrow short and lend long. This entails potential dangers of illiquidity at the maturity of the deposits collected in order to finance the loans. One aspect of the liquidity risk is covered by the roll-over mechanism, through which the rate of interest on the lending operations is mechanically connected with the London Interbank Offered Rate (LIBOR), so that the extra price for money is transferred to the borrower. For small and medium sized banks and banks without easy access to dollars it is not always certain that there will always be an adequate supply of funds.

Country risks, risks from Government action and liquidity risks may all end in a liquidity crisis leading to defaults on a scale that would challenge the stability of the West's banking system leading to the possibility of flights from paper money into real assets. Real assets, particularly those with the qualities possessed by gold, would greatly benefit from this collapse of the international monetary system.

The Role of OPEC

With this background in mind it is instructive to examine the size and deployment of OPEC oil surpluses. OPEC have moved to a pricing policy designed to offset the effects of inflation and fluctuations in the dollar exchange rate. The sharp increase in oil prices during 1979 and 1980 have resulted in a massive transfer of resources to the oil producing countries. As a result the aggregate OPEC current account surplus totalled $96 billion in 1980 and has been estimated by Morgan Guaranty Trust Company at $110 billion for 1981 despite dramatic increases in domestic spending in individual oil producing countries. As the oil producers accumulate large financial surpluses in 1980, 1981 and beyond, the investment decisions of OPEC financial managers will have an important effect on the gold market as well as on worldwide exchange rates and interest rates.

By the end of 1981 OPEC's net external assets are expected to have increased from the 1979 level of almost $ 240 billion to as much as $ 445 billion. In the past, these investments have roughly been split between short- and long-term assets. Bank of England data (*see Table 8.1*) indicate that the $ 236 billion aggregate investment position accumulated between 1974—79 included $ 115 billion in bank deposits and $ 17 billion in US or British government securities. An estimated 68 per cent of the bank deposits were dollar-denominated; of this, some $ 14 billion was held in the USA and $ 64 billion in the Euromarket. The remaining $ 37 billion in non-dollar OPEC bank deposits were held mostly in Deutschemarks, Swiss francs, and Japanese yen. Overall, roughly three-quarters of OPEC's net foreign asset position was estimated to be in dollar-denominated instruments.

Table 8.1. OPEC Investments (end of 1979, billions of dollars).

Bank deposits		115
Euromarkets	89	
Domestic markets	26	
US/UK Government Securities		17
Short term	7	
Long term	10	
Portfolio and Direct Investments*		58
IMF and World Bank		8
Loans to Developing Countries**		38
Total		236

*Includes loans and other items.
**Includes subscriptions to regional and development
  agencies.
Source: Bank of England.

The surplus funds accumulated during 1980 and 1981 are also likely to be invested heavily in dollar assets and, initially, in the form of short-term bank deposits. Recent data suggest that the oil producing countries have a growing preference to hold these assets outside of the USA and may be inclined to invest a greater share in non-dollar investments than in the past.

Forecasts of individual country external positions, upon which these aggregate forecasts are based, highlight the concentration of current account surpluses among four or five countries (*see Table 8.2*). Saudi Arabia, Kuwait, Iraq, and Libya are expected to account for about three-quarters of the aggregate surplus both in 1980 and 1981.

Table 8.2. OPEC Current Account Surpluses
(Billions of dollars).

	1979[e]	1980[e]	1981[e]
Saudi Arabia	15.5	41.0	32.0
Kuwait	13.9	19.3	20.4
Iraq	11.4	14.8	11.7
United Arab Emirates	4.9	8.4	8.0
Iran	4.5	3.4	11.0
Libya	4.2	14.2	9.3
Nigeria	1.7	7.5	5.4
Qatar	1.7	3.6	3.8
Algeria	−1.5	2.3	2.0
Indonesia	0.9	0.8	−1.7
Gabon	0.4	0.6	0.3
Ecuador	−0.6	−0.5	−0.6
Venezuela	−1.9	0.0	−0.6
Total	$55.1	$115.4	$101.1

[e] = estimate
Source: First National Bank of Chicago.

In Saudi Arabia, export revenues are forecast to reach $ 94 billion in 1980 and $ 96 billion in 1981 (the result of higher oil prices and lower output in 1981). Investment income is expected to approach $ 11 billion in both years (*see Table 8.3*). If these forecasts are accurate, Saudi Arabia's external assets, which include the assets of the Saudi Arabian Monetary Authority, should approach $ 150 billion by the end of 1981, double the 1979 level (*see Table 8.4*).

As already discussed in earlier chapters gold is the only asset which can be transported to a safe place with relative ease, which is marketable all over the world and is beyond the direct control of national monetary authorities. The OPEC countries have both the ability and the willingness to buy gold. The ability comes from their large surpluses and continued

Table 8.3. Interest Income (billions of dollars).

	1979[e]	1980[e]	1981[e]
Saudi Arabia	6.9	11.6	10.7
Kuwait	4.0	6.6	6.3
Iraq	2.0	4.0	3.7
U.A.E.	1.6	2.6	2.4
Libya	1.0	2.3	2.4

[e] = estimate
Source: First National Bank of Chicago.

Table 8.4. Net Foreign Assets*
(billions of dollars).

	1979[e]	1980[e]	1981[e]
Saudi Arabia	$77	$118	$150
Kuwait	45	64	84
Iraq	26	41	52
UAE	18	26	34
Libya	12	26	35

*At year-end
[e] = estimate
Source: First National Bank of Chicago.

investment income illustrated in Tables 8.1 to 8.4. The willingness stems from their perception of the aforementioned characteristic of gold combined with disillusionment about the alternatives. The large dependence on dollar denominated instruments mentioned above combined with the US blockage of Iranian assets has given the Middle East countries reason to diversify out of the dollar into some other asset. With such a large investment portfolio it only requires a small proportion of this to move into gold for there to be an enormous impact on the gold price. In addition political instability in the Middle East, witness the Mecca riots and the Iranian–Iraqi war, makes the leaders in these countries anxious to acquire assets with the above mentioned characteristics. These considerations induced the Libyan Government to increase its gold reserves by some 32 per cent (to 110 tonnes) between September 1980 and April 1981.

The demand for gold from the Middle East comes as no surprise when one examines movements in the terms of trade of OPEC countries. As can be seen from Table 8.5 below, from early 1974 to early 1978 there was no change in the terms of trade for the OPEC countries. The terms of trade is a measure of what a country receives for its exports and what it pays for its imports. The September 1973 oil price rise, highlighted by the figure of

Table 8.5. Terms of Trade Developments 1962–1979
(percentage changes).

	Average 1962–72	1973	1974	1975	1976	1977	1978	1979
Oil exporting countries	1	15	137	−5	4½	1	−10½	28

Source: IMF Annual Report 1980.

137 in Table 8.5, was inflated away by the West in later years. As was discussed in Chapter 4, there is a certain logic in oil producers wishing to exchange the finite, non-renewable asset of oil by another finite, non-renewable asset gold, particularly if this process also imposes monetary discipline on the creators of paper currency in the countries to whom they are exporting.

The Role of Central Banks

In the framework of the original Bretton Woods system, the main purpose of foreign exchange reserves was in fact to provide national residents, through their commercial banks, with the necessary amount of foreign cash needed to meet obligations *vis-a-vis* foreign exporters or suppliers of overseas services. In most cases the central bank was the only institution authorised, subject to exchange controls systems of individual countries. to hold sizable amounts of foreign currencies and their reserves largely corresponded to the working balances needed by the countries concerned in order to sustain a given level of external trade. The subsequent liberalisation of exchange control regulations, the increased liquidity position of central banks, the enlarged credit facilities provided by Euromarkets and the international activities of national banking systems, as well as the abolition of fixed parities, have relieved central banks from their duty as provider of last resort of external reserves. The need, however, for the maintenance of reserves by central banks has not disappeared. The liquidity or usability of gold in case of need is no longer hampered by an unrealistic official price which actually froze gold holdings in central banks vaults.

As was mentioned in Chapter 5, central banks have perceived the benefits of gold holdings in times of crisis. The numerous examples of this, detailed below, have meant that central banks are always anxious, where possible, to increase their gold holdings. Italy and Portugal in 1974 and 1976 were able to obtain badly needed foreign exchange loans by pledging part of their gold reserves. Also Latin American countries received foreign currency loans against gold as collateral. The Banque de France started in 1976 to appreciate the value of its gold reserves, in order to compensate for the diminution of its currency reserves caused by balance of payments deficits. During 1977 and 1978 South Africa only escaped a stringent western imposed financial policy by swapping part of their gold reserves for foreign exchange. Even European countries with strong currencies such as the Federal Republic of Germany, Switzerland and Austria, all of a sudden at the end of 1978 remembered their gold holdings. To overcome or at least to compensate part of their losses generated by the depreciation of their US dollar holdings during 1978, these countries appreciated the book value of their gold holdings. Exceptions to this increase in gold holdings do occur as when the central bank of Costa Rica sold $42 million of reserves in May 1981.

A further factor suggesting greater future stability in the gold price, widely discussed in earlier chapters, is the formation of the European Monetary Fund and its unit of account, the European Currency Unit. In this respect the mobilisation of central bank gold reserves will make gold, in the words of stockbrokers Laing and Cruickshank 'the sun around which adjustments in parities will be made'. Thus central banks will be concerned to act in the free gold market to ensure a relatively stable price for bullion.

The Laffer Plan

Increasing worldwide distrust of paper currencies has resulted in widespread interest in restoring gold related standard which would restrict the ability of governments to increase public expenditure without adequate funding. The merits of the gold standard were discussed in detail in Chapter 3. In this Section the details of two widely discussed plans, the Laffer Plan and the Bareau Plan are outlined. These plans have acquired increased importance following the election of Ronald Reagan as President of the USA. Reagan is a sound money man and has a strong regard for the discipline of a gold standard.

Professor Arthur Laffer, one of Mr. Reagan's closest pre-election advisers has put forward a plan to restore confidence in the dollar. Laffer argues that the decline of the dollar's value in world markets has become an engine of inflation in its own right, forcing US consumers to continually pay more for imported goods, and only a return to a programme of convertibility of the dollar into gold would put things rights again. Laffer believes in preserving what he calls the 'moneyness of money', ie that a currency's purchasing power be maintained from one day to the next, from one year to the next and so on, so that people will want to hold it, to use it, to work for it. If prices were expected to be stable over the foreseeable future there would then be a dramatic shift out of gold into dollars. Thus gold convertibility, combined with appropriate economic policies, would encourage holders of dollars to keep them thereby exerting downward pressure on the gold price.

Laffer, using illustrative specific values, has set out his plan for linking gold to the dollar. The following steps would be announced:

(1)   The USA would announce its intention of returning to a convertible dollar — that is, to a dollar both internally and externally convertible into gold at a new official price — at some specified time in the future.

(2)   At the end of whatever transition period is announced, the Federal Reserve would establish an official dollar parity in terms of gold which would be based on the prevailing market price for gold.

(3)   From this point, the Federal Reserve would stand ready to buy and sell gold freely at the official price, subject to a small transactions charge.

(4)   The Federal Reserve would have a target dollar value of gold reserves equal to, say, 40 per cent of the dollar value of its liabilities. There would also be an upper reserve limit of, say, 70 per cent, and a lower reserve limit of, say, 10 per cent; and within these limits, there would be a narrower reserve band of, say, 50 per cent and 30 per cent.

(5)   Within the narrower reserve band, the Federal Reserve would be permitted to follow a discretionary monetary policy; but, if the reserve ratio should move beyond these limits, monetary policy would be subject to statutory rules, with the monetary base being expanded or contracted as need be, according to some specified formula. In other words, an automatic adjustment mechanism (along pre-1914 gold standard lines) would be set in motion, once the allowable monetary tolerances had been breached.

(6)   If, despite the mandated adjustment mechanism, the outer reserve band should nevertheless be breached, the dollar's convertibility would be suspended temporarily, and the dollar price of gold would be permitted to find its new market level.

During the subsequent adjustment period, the Federal Reserve Board's discretion would be removed. The monetary authority would be required to run policies such that the monetary base would experience no growth. At the end of this adjustment period the dollar's gold parity would be restated at the then prevailing market level.

Laffer's proposal attempts to remedy two serious defects inherent in most systems to return to gold convertibility. The original fixing price of gold no longer would be left to the vicissitudes of political pressure. This would avoid the necessity of making the overall economy adjust to some inappropriate price of gold. The second criticism, to which Laffer's Plan provides an answer, is as to what happens if there is an explicit change in the market for gold itself. If gold became excessively plentiful or scarce due to conditions beyond the control of the monetary authority the domestic economy would be insulated from this by breaking the gold convertibility. Examples of the above possibilities would be significant new gold finds or a massive increase in the demand for gold.

Laffer's Plan is somewhat complicated by the politically independent US Federal Reserve. As the Federal Reserve would be being asked to increase and decrease the money supply at different times its co-operation would be essential for Laffer's plan to be implemented. However, Laffer replies to this problem by pointing out that Paul Volcker, who will probably be the Chairman of the Federal Reserve until 1987, has long argued in favour of gold convertibility.

Various consequences of Laffer's plan would occur. With the dollar being as good as gold there would be a reduction of inflationary expectations throughout the economy benefitting consumers, savers and investors alike. In addition the foreign exchange value of the dollar would rise. In order to offset the reduction in their own exchange rates non-US central banks would be likely to sell dollars in exchange for their own domestic monies. This move, in effect a return to fixing exchange rates, would result in their own currencies in turn being effectively linked to gold.

The Bareau Plan

As discussed in Chapter 3 there can be no question of going back to the classic gold standard of old, to the clink of gold coins in our pockets, even to the ability of private individuals and corporations to go to their respective central banks and exchange their domestic currencies into gold bars as occurred under the gold bullion standard. With the deficiencies from the investor's points of view inherent in investment in domestic monies, such as the dollar combined with the limited usage of the SDR and the breakdown of discussions about the substitution account the way, it is argued, is open for a new initiative on the international monetary front.

Paul Bareau has put forward an interesting plan which attempts to obtain the benefits without the defects of the gold standard as a possible means of officially remonetising gold for the 1980s. (By remonetisation is meant a return to an official gold price.) What he proposes is the development of a multiple reserve currency standard, each reserve currency the centre of a bloc within which reasonable stability would prevail, and between which the authorities should strive to achieve the maximum degree of stability. This system in turn would have far greater promise of stability and endurance and far greater immunity from the temptations of inflation if the major currencies command some measure of conditional restricted convertibility into gold.

The re-emergence of official prices for gold would inevitably, as with the Laffer plan, involve official intervention in the free gold market. However, Bareau stresses that there should be strenuous efforts made to avoid the costly errors of the gold pool of the 1960s, described in Chapter 4. Thus the official price must be flexible and not fixed. An official gold price would impose external discipline on governments, thereby reducing the possibilities of lax monetary policies. The IMF would have the power, as it had under the original Bretton Woods Articles of Agreement, to make across the board changes in the price of gold in terms of all member currencies. There would thus be a system of flexible exchange parities. These powers should be actively used. In the appropriate circumstances the required flexibility would be there, but in addition there would be the discipline of gold parities and exchange parities. A change in these parities would. call for a measure of moral courage and deterrent publicity, Bareau argues, far greater than is required of the present regime of floating and inconvertibility.

The convertibility into gold would be restricted to inter-central banking, inter-Governmental transactions. But it would have to be real not symbolic if it was to have its disciplinary effect. This new system would require official intervention in the free gold market. It would remove the volatility, the essentially speculative froth in the market, though it would not deter the intelligent operator from trying to anticipate the next shift in this 'crawling peg' official price of gold. It would make for a much more stable gold-market not only because of this stabilising official intervention, but by removing from the free market the ever present threat of an avalanche of demonetised, official gold. It would, Bareau claims, combine freedom and flexibility with rediscovery of some of the discipline which was lost with the breakdown of the gold standard in the 1930s and later with the breakdown of the Bretton Woods gold-dollar standard in 1971.

A Euro Gold Market?

Swiss francs, US dollars and Deutschemarks all share one defect as international monies: their base is national and their values depend on sound monetary policies being followed at home. All are subject to the risk of imposition of exchange restrictions by the issuer. International investors who have often reason to be sceptical about the politics of money have a natural inclination towards one which is not a national brand. It is for this reason that gold has emerged and is likely to remain as the ultimate in stateless money.

Whilst gold continues to float against national currencies gold remonetisation is likely to be evidenced by the birth of credit markets denominated in gold with gold thus reasserting its role as a unit of account. Brendan Brown has illustrated in his book, *Money, Hard and Soft* how such a credit market would work. Interest rates on top rated gold bonds would be set at a level sufficient to compensate their holders for the disadvantage of holding gold paper rather than the yellow metal itself. Investors trading interest in gold could then be concentrated in the gold paper market.

Investment in bullion gold would then only be necessary as a form of investment against future uncertainties. The interest rate of the gold paper would have to reflect the investment versus the insurance gains of holding gold in this form. The creation of a credit market in gold paper would be likely to decrease the volatility of the gold price. Heavy demand, for what, effectively would be Euro-gold, would be absorbed in part by its interest yield falling: pressure on the spot gold price would in consequence be less. When investors shifted from Euro-gold into other euro-monies, its yield would rise: gold borrowers would be discouraged, hence reducing the fall of the gold price.

A large Euro-gold market would provide the best protection against gold price volatility being induced by investment inflows and outflows. Investors and borrowers who used gold as a unit of account, or a component of their basket unit of account, would increase their issue of gold paper when gold interest rates fell and conversely. Further, the larger are net private holdings of Euro-gold the smaller are likely to be proportionate shifts in net stock demand. Thus a smaller change in the gold price will restore equilibrium in the Euro-gold market.

In January 1981 Drexel Burnham Lambert announced the issue of a gold-linked bond for Refinement the US recycler of precious metals. Altogether 10,000 15 year bonds have been issued with a nominal value of 10 ounces of gold each. The bonds bear interest at a nominal rate of between 4 and 5 per cent. Interest and principal are payable in gold to the extent that they are multiples of 100 ounces. Otherwise, payments will be in US dollars based on an average market price of gold for a specified period preceding the payment date.

European governments are other likely sponsors of a Euro-gold market. Many hold large gold reserves, hence their issues of gold bonds would be hedged many times over. European institutions such as the European Investment Bank, The European Coal and Steel Community and the EEC itself could be among the first to make gold bond issues. Most EEC members with the exception of the UK and West Germany, have never wholeheartedly supported the USA in its anti-gold campaign. Some LDC governments and other sovereign powers with gold reserves would soon follow the European example. A gold denominated syndicated credit market would be a possible further development.

Election of President Reagan:
Implications for the Gold Market

The change in US presidency from Jimmy Carter to Ronald Reagan and the switch from a Democrat to a Republican administration will affect the gold market as the country's policy regarding gold switches from one of demonetisation to remonetisation. As normally occurs after an election there may well be a marked difference between some of Reagan's policies as promised during his electioneering and those actually put into practice. The Republican party's monetary sub-committee only just stopped short of calling for a return to the gold standard. Senator William Roth clarified Republican policy in this regard by saying 'Our open mindedness about a gold standard indicates our displeasure with the Federal Reserve for ignoring general price levels in setting monetary policy and printing too much money. The way to stop inflation is for the Federal Reserve to stop printing this money and pursue a rate of monetary growth in line with the rate of expansion in the economy.' Certainly, if the Republicans are to succeed in restoring what the economist Arthur Laffer terms the 'moneyness of money' the need for a gold standard would be less anyway. A well managed dollar will at least restore confidence amongst dollar holders and, when this is associated with lower inflation rates, the huge investment and hedge purchasing of gold and paper gold may decline significantly.

There are, however, some contradictions in Reagan's policies which cast doubt on his ability to restore long term worldwide confidence in the dollar. It is unclear how he will reconcile his preference for free-market solutions with reassurances of concern given during the campaign to producer and worker groups that want to retain or secure protection from foreign competition. More importantly Reagan faces a formidable challenge in achieving budgetary restraint while at the same time accommodating Kemp-Roth income tax reductions and increased military expenditure. The Kemp-Roth Bill aims to reduce the marginal tax rate for individuals by 30 per cent, spread over three years. Despite these conflicting objectives Reagan aims to achieve a balanced Federal Budget by fiscal 1983. The odds of achieving this are somewhat slight. Over the short run, however, as was seen by the behaviour of the gold market in early 1981, the introduction of tight money as evidenced by high interest rates is bearish for the gold market. The reasons for this are three fold. Firstly, gold bought on margin in Europe becomes rather expensive to hold when interest rates are high. Secondly, operators on the North American futures markets are encouraged to short gold at high 'premium prices' compared to quotations in the spot market. Thirdly, alternative assets, particularly in the fixed interest area, look more attractive. Over the longer run rising inflation will be bullish for gold and in addition the US may soon join the majority of other countries in valuing its gold at market related prices thus giving support to gold's monetary role.

In 1980, the US prime rate swung from 20 per cent in April to 10 per cent in July and to above 20 per cent for the first half of 1981. Such a high rate is clearly depressing for the world economy at large (US production represents around 40 per cent of the global production of the industrial world) not so much because of its level in nominal terms but because of a very significant divergence from the rate of inflation. Throughout much of 1980 and 1981 the USA has had a high real rate of interest higher than in most industrial countries. This high real rate of interest has brought down the gold price. Private sector gold demand is only likely to pick up again once US interest rates demonstrate a clear downward trend. US interest rates incorporate a very large inflation premium. Consequently only when the US inflation rate falls will the gold price start to rise. Information on expected inflation rates can be inferred from the mid-1981 economic review provided by the Reagan administration. This forecasts a lowering of the inflation rate, which can then be paralleled by falls in interest rates, taking place from the beginning of 1981 onwards.

Conclusion

As Professor Triffin of Yale succinctly put it,

> 'Nobody could ever have conceived of a more absurd waste of human resources than to dig gold in distant corners of the earth for the sole purpose of transporting it and reburying it immediately afterwards in other deep holes, especially excavated to receive it and heavily guarded to protect it.'

Despite this view, for the various reasons outlined throughout this book, there has been a significant structural change in the gold market. For both governments and investors alike gold has acquired a renewed importance. Central banks worldwide are revaluing their gold reserves; illustrating that gold is firmly back in the centre of the international monetary system. For private investors both in the Middle East as well as in Europe and the USA gold is becoming an accepted asset as part of investors' portfolios. The US Congress has appointed a Committee to make recommendations on the use of gold for domestic and international monetary purposes. It is scheduled to submit its report towards the end of 1981. Should it recommend a revaluation of the US gold reserves the way would be open for a resurgence of interest in gold. Indeed the US authorities would have a vested interest in intervening in the gold market to stabilise the price. If US interest picks up it would only require a small proportion of the estimated $350 billion US pension monies to move into gold to send gold back into the stratosphere. In the language of economics gold has a high elasticity of expectations, that is, the ratio of expected price increases to present price increases is high. With the speculative motive feeding on itself gold would no longer be perceived of as a 'barbarous relic' with its importance in the future being assured.

Appendix
MINTED GOLD COINS

*Reproduced from *How to Invest in Gold* by Peter C. Cavelti published in Canada by Maximus Press Limited, Toronto and in the USA by Follett Publishing Corporation, Chicago, by permission of the author.

Country or Coin Area		Gold Content in oz.	Years Produced
Afghanistan	1 Tilla	0.133	1896-1919
	5 Amani	0.638	1921
	2 Amani	0.263	1921-1924
	1 Amani	0.132	1919-1932
	1/2 Amani	0.066	1921-1928
	21/2 Amani	0.434	1925-1928
	1 Amani	0.173	1925-1928
	1/2 Amani	0.087	1925-1928
	1 Habibi	0.122	1929
	20 Afghani	0.173	1929-1930
Africa	3 Dinars	0.395	660-1902
	2 Dinars	0.263	660-1902
	1 Dinar	0.132	660-1902
	1/2 Dinar	0.066	660-1902
	1/4 Dinar	0.033	660-1902
Ajman	100 Riyals	0.627	1970
	75 Riyals	0.449	1970
	50 Riyals	0.299	1970
	25 Riyals	0.149	1970
Albania	100 Francs	0.933	1926-1938
	50 Francs	0.467	1938
	20 Francs	0.187	1926-1938
	10 Francs	0.093	1927
	500 Leks	2.857	1968-1970
	200 Leks	1.142	1968-1970
	100 Leks	0.571	1968-1970
	50 Leks	0.285	1968-1970
	20 Leks	0.114	1968-1970
Anguilla	100 Dollars	1.428	1967
	20 Dollars	0.285	1967
	10 Dollars	0.142	1967
	5 Dollars	0.071	1967
Argentina	5 Pesos	0.233	1881-1896
	21/2 Pesos	0.117	1881-1884
Asia	5 Mohurs	1.718	1200-1947
	2 Mohurs	0.687	1200-1947
	1 Mohur	0.344	1200-1947
	1/2 Mohur	0.172	1200-1947

Country or Coin Area		Gold Content in oz.	Years Produced
Asia (cont.)	1/4 Mohur	0.086	1200-1947
	1/8 Mohur	0.043	1200-1947
	1/16 Mohur	0.021	1200-1947
	1/32 Mohur	0.011	1200-1947
	3 Dinars	0.395	660-1902
	2 Dinars	0.263	660-1902
	1 Dinar	0.132	660-1902
	1/2 Dinar	0.066	660-1902
	1/4 Dinar	0.033	660-1902
Austria	4 Ducats	0.443	1793-1915
	2 Ducats	0.221	1799-1804
	1 Ducat	0.111	1792-1915
	1 Soverain	0.327	1781-1800
	1/2 Soverain	0.163	1781-1800
	1 Sovrano	0.328	1820-1856
	1/2 Sovrano	0.164	1820-1856
	1 Krone	0.321	1858-1866
	1/2 Krone	0.161	1858-1866
	8 Florins	0.187	1870-1892
	4 Florins	0.093	1870-1892
	100 Corona	0.980	1908-1915
	20 Corona	0.196	1892-1912
	10 Corona	0.098	1892-1912
	100 Kronen	0.980	1923-1924
	20 Kronen	0.196	1923-1924
	100 Schilling	0.681	1926-1938
	25 Schilling	0.170	1926-1938
Bahamas	100 Dollars	1.177	1967-1971
	50 Dollars	0.588	1967-1971
	20 Dollars	0.235	1967-1971
	10 Dollars	0.116	1967-1971
	100 Dollars	0.941	1972
	50 Dollars	0.470	1972
	20 Dollars	0.188	1972
	10 Dollars	0.094	1972
	100 Dollars	0.273	1973
	50 Dollars (Crawfish)	0.136	1973
	50 Dollars (Flamingo)	0.294	1973
	20 Dollars	0.054	1973
	10 Dollars	0.027	1973

Country or Coin Area		Gold Content in oz.	Years Produced
Bahamas	200 Dollars	0.321	1974-1975
(cont.)	150 Dollars	0.241	1974-1975
	100 Dollars (Independence)	0.289	1974-1975
	100 Dollars (4 Flamingoes)	0.160	1974-1975
	50 Dollars	0.080	1974-1975
Bahrain	10 Dinars	0.471	1968
Barbados	100 Dollars	0.099	1975
Belgium	100 Francs	0.933	1853-1912
	40 Francs	0.373	1834-1841
	25 Francs	0.233	1847-1850
	20 Francs	0.187	1835-1914
	10 Francs	0.093	1849-1912
Bermuda	20 Dollars	0.235	1970
	100 Dollars	0.226	1975
Bhutan	5 Sertums	1.178	1966
	2 Sertums	0.471	1966
	1 Sertum	0.235	1966-1970
Biafra	25 Pounds	2.354	1969
	10 Pounds	1.177	1969
	5 Pounds	0.470	1969
	2 Pounds	0.235	1969
	1 Pound	0.117	1969
Bolivia	1 Onza	1.041	1868
	35 Grams	1.125	1952
	14 Grams	0.450	1952
	7 Grams	0.225	1952
	3½ Grams	0.113	1952
Botswana	10 Thebe	0.327	1966
	150 Pulce	0.471	1976
Brazil	4000 Reis	0.241	1695-1833
	2000 Reis	0.120	1695-1793
	1000 Reis	0.060	1695-1787
	20000 Reis	1.579	1724-1727

Country or Coin Area		Gold Content in oz.	Years Produced
Brazil (cont.)	10000 Reis	0.789	1724-1727
	4000 Reis	0.315	1703-1727
	2000 Reis	0.158	1703-1727
	1000 Reis	0.079	1708-1727
	400 Reis	0.031	1725-1730
	12800 Reis	0.842	1727-1733
	6400 Reis	0.421	1727-1833
	3200 Reis	0.210	1727-1786
	1600 Reis	0.105	1727-1784
	800 Reis	0.052	1727-1786
	400 Reis	0.026	1730-1734
	20000 Reis	0.528	1849-1922
	10000 Reis	0.264	1849-1922
	5000 Reis	0.132	1854-1859
	300 Cruzeiros	0.492	1972
British	5 Pounds	1.177	1817-1964
Empire	2 Pounds	0.471	1817-1964
	1 Pound	0.235	1817-1964
	$1/2$ Pound	0.118	1817-1964
British Virgin Islands	100 Dollars	0.205	1975
Brunei	$1,000	1.472	1978
Bulgaria	100 Leva	0.933	1894-1912
	20 Leva	0.186	1894-1912
	10 Leva	0.093	1894
	20 Leva	0.488	1963-1964
	10 Leva	0.244	1963-1964
Burma	5 Rupees	0.101	1880
	4 Rupees	0.081	1852-1878
	2 Rupees	0.040	1852-1880
	1 Rupee	0.020	1852-1880
Burundi	100 Francs	0.925	1962
	50 Francs	0.463	1962
	25 Francs	0.2315	1962
	10 Francs	0.092	1962
	100 Francs	0.868	1965
	50 Francs	0.434	1965
	25 Francs	0.217	1965
	10 Francs	0.086	1965

Country or Coin Area		Gold Content in oz.	Years Produced
Cameroun	20000 Francs	2.025	1970
	10000 Francs	1.013	1970
	5000 Francs	0.506	1970
	3000 Francs	0.303	1970
	1000 Francs	0.101	1970
Canada	5 Dollars	0.241	1912-1914
	10 Dollars	0.483	1912-1914
	20 Dollars	0.538	1967
	100 Dollars Olympic	0.250	1976
	100 Dollars Olympic	0.500	1976
	100 Dollars Jubilee	0.500	1977
	100 Dollars Unity	0.500	1978
	100 Dollars Year of Child	0.500	1979
	50 Dollars Maple Leaf	1.000	1979-
Cayman Islands	25 Dollars	0.253	1972
	100 Dollars	0.364	1974-1975
Central Africa	20000 Francs	2.025	1970
	10000 Francs	1.013	1970
	5000 Francs	0.506	1970
	3000 Francs	0.303	1970
	1000 Francs	0.101	1970
Central America	50 Pesos	0.578	1970
Chad	20000 Francs	2.025	1970
	10000 Francs	1.013	1970
	5000 Francs	0.506	1970
	3000 Francs	0.303	1970
	1000 Francs	0.101	1970
Chile	10 Pesos	0.439	1853-1890
	5 Pesos	0.219	1858-1873
	2 Pesos	0.088	1857-1875
	1 Peso	0.044	1860-1873

Country or Coin Area		Gold Content in oz.	Years Produced
Chile (cont.)	20 Pesos	0.352	1896-1917
	10 Pesos	0.176	1895-1901
	5 Pesos	0.088	1895-1900
	100 Pesos	0.588	1926-1963
	50 Pesos	0.294	1926-1962
	20 Pesos	0.117	1926-1961
	500 Pesos	2.941	1967
	200 Pesos	1.176	1967
	100 Pesos	0.588	1967
	50 Pesos	0.294	1967
China, Republic	20 Dollars	0.428	1919
	10 Dollars	0.214	1916-1919
China, Yunnan	10 Dollars	0.260	1919
	5 Dollars	0.130	1919
China, Nationalist	2000 Yuan	0.868	1965
	1000 Yuan	0.434	1965
	2000 Yuan	0.949	1966
Colombia	20 Pesos	0.933	1859-1877
	10 Pesos	0.467	1856-1877
	5 Pesos	0.233	1856-1885
	2 Pesos	0.098	1856-1876
	1 Peso	0.047	1856-1878
	1 Peso	0.471	1919-1924
	5 Pesos	0.235	1913-1930
	2$1/2$ Pesos	0.118	1913-1928
	1500 Pesos	1.867	1968-1971
	500 Pesos	0.622	1968-1971
	300 Pesos	0.373	1968-1971
	200 Pesos	0.249	1968-1971
	100 Pesos	0.124	1968-1971
	1500 Pesos (Bank)	0.553	1973
	1500 Pesos (Valencia)	0.249	1973
	2000 Pesos	0.373	1973
	1000 Pesos	0.124	1973-1975
	1000 Pesos	0.249	1975
Costa Rica	10 Pesos	0.402	1870-1876
	5 Pesos	0.201	1867-1875

Country or Coin Area		Gold Content in oz.	Years Produced
Costa Rica	2 Pesos	0.081	1866-1876
(cont.)	1 Peso	0.040	1864-1872
	20 Colones	0.450	1897-1900
	10 Colones	0.225	1897-1900
	5 Colones	0.113	1899-1900
	2 Colones	0.045	1897-1928
	1000 Colones	5.615	1970
	500 Colones	2.156	1970
	200 Colones	0.862	1970
	100 Colones	0.431	1970
	50 Colones	0.216	1970
	1500 Colones	0.968	1974
Croatia	500 Kuna	0.282	1941
Cuba	20 Pesos	0.967	1915-1916
	10 Pesos	0.484	1915-1916
	5 Pesos	0.242	1915-1916
	4 Pesos	0.193	1915-1916
	2 Pesos	0.097	1915-1916
	1 Peso	0.048	1915-1916
Czechoslovakia	10 Ducats	1.106	1929-1951
	5 Ducats	0.553	1929-1951
	5 Ducats (Commemorative)	0.634	1929
	4 Ducats	0.443	1928
	3 Ducats	0.380	1929
	2 Ducats	0.221	1923-1951
	1 Ducat	0.111	1923-1951
	1 Ducat (Commemorative)	0.127	1929
Dahomey	25000 Francs	2.572	1970
	10000 Francs	1.029	1970
	4000 Francs	0.514	1970
	2500 Francs	0.257	1970
Danish West Indies	10 Daler	0.467	1904
	4 Daler	0.187	1904-1905
Danzig	25 Gulden	0.235	1923-1930
Denmark	2 Christian d'or	0.383	1826-1870

Country or Coin Area		Gold Content in oz.	Years Produced
Denmark (cont.)	1 Christian d'or	0.191	1775-1869
	1 Courant Ducat	0.088	1771-1785
	1 Ducat	0.111	1771-1802
	10 Daler	0.467	1904
	4 Daler	0.187	1904-1905
	20 Kroner	0.259	1873-1931
	10 Kroner	0.130	1873-1917
Dominican Republic	30 Pesos	0.857	1955
	20 Pesos	0.338	1974
Ecuador	10 Sucres	0.235	1899-1900
	1 Condor	0.242	1926
Egypt	500 Piastres	1.196	1861-1960
	100 Piastres	0.239	1839-1960
	50 Piastres	0.120	1839-1958
	20 Piastres	0.048	1923-1938
	10 Piastres	0.024	1839-1909
	5 Piastres	0.012	1839-1909
	10 pounds	1.463	1964
	5 Pounds	0.731	1964
	1 Pound	0.239	1955-1960
	1 Pound	0.225	1970-1973
	$1/2$ Pound	0.120	1958
Ethiopia	2 Wark (± 14 grams)	0.405	1889-1917
	1 Wark (± 7 grams)	0.203	1889-1931
	$1/2$ Wark (± 3.5 grams)	0.101	1889-1931
	$1/4$ Wark (± 1.75 grams)	0.051	1889
	$1/8$ Wark (± 0.875 grams)	0.025	1889
	200 Dollars	2.315	1966
	100 Dollars	1.157	1966-1972
	50 Dollars	0.579	1966-1972
	20 Dollars	0.232	1966
	10 Dollars	0.116	1966
Europe	100 Ducats	11.063	1280-1960
	50 Ducats	5.532	1280-1960

Country or Coin Area		Gold Content in oz.	Years Produced
Europe (cont.)	20 Ducats	2.213	1280-1960
	10 Ducats	1.106	1280-1960
	5 Ducats	0.553	1280-1960
	4 Ducats	0.443	1280-1960
	3 Ducats	0.332	1280-1960
	2 Ducats	0.221	1280-1960
	1 Ducat	0.111	1280-1960
	1/2 Ducat	0.055	1280-1960
	1/4 Ducat	0.028	1280-1960
	1/8 Ducat	0.014	1280-1960
	1/16 Ducat	0.007	1280-1960
	1/32 Ducat	0.003	1280-1960
Falkland	5 Pounds	1.176	1974
Islands	2 Pounds	0.471	1974
	Sovereign	0.0235	1974
	1/2 Sovereign	0.118	1974
Fiji	100 Dollars	0.504	1974
Finland	20 Markka	0.187	1878-1913
	10 Markka	0.093	1878-1913
	200 Markka	0.244	1926
	100 Markka	0.122	1926

France	*The weight of French coins prior to 1803 fluctuated greatly.* *The gold contents and values quoted reflect averages.*		
	1 Ecu d'or	0.105	1266-1641
	1 Chaise d'or	0.151	1285-1422
	1 Royal d'or	0.135	1285-1461
	1 Lion d'or	0.158	1328-1350
	1 Pavillion d'or	0.164	1328-1350
	1 Ange d'or	0.233	1328-1350
	1 Mouton d'or	0.151	1350-1422
	1 Franc à cheval	0.125	1350-1461
	1 Franc à pied	0.123	1350-1380
	1 Salut d'or	0.125	1380-1461
	1 Heaume d'or	0.150	1380-1422
	1 Henry d'or	0.111	1550-1559
	2 Louis d'or	0.450	1640-1792
	1 Louis d'or	0.222	1640-1793
	1/2 Louis d'or	0.111	1640-1784
	100 francs	0.933	1855-1913

Country or Coin Area		Gold Content in oz.	Years Produced
France (cont.)	50 francs	0.467	1855-1904
	40 francs	0.373	1803-1839
	20 francs	0.187	1803-1914
	10 francs	0.093	1854-1914
	5 francs	0.047	1854-1889
	100 francs	0.190	1929-1936
Fujairah	200 Ryials	1.200	1969
	100 Ryials	0.600	1969-1971
	50 Ryials	0.300	1970
	25 Ryials	0.150	1970
Gabon	100 Francs	0.926	1960
	50 Francs	0.463	1960
	25 Francs	0.232	1960
	10 Francs	0.093	1960
	20,000 Francs	2.025	1969
	10,000 Francs	1.013	1969
	5,000 Francs	0.506	1969
	3,000 Francs	0.304	1969
	1,000 Francs	0.101	1969
German East Africa	15 Rupees	0.217	1916
German New Guinea	20 Marks	0.231	1895
	10 Marks	0.115	1895
Germany	10 Taler	0.385	1742-1857
	5 Taler	0.192	1699-1856
	2$^{1}/_{2}$ Taler	0.096	1699-1855
	1 Pistole	0.192	1699-1856
	1 Frederick d'or	0.192	1699-1856
	1 Carolin	0.240	1726-1782
	$^{1}/_{2}$ Carolin	0.120	1726-1737
	$^{1}/_{4}$ Carolin	0.060	1726-1736
	4 Ducats	0.443	1797-1844
	2 Ducats	0.221	1768-1811
	1 Ducat	0.111	1764-1872
	2 Taler	0.077	1792-1830
	40 Francs	0.373	1813
	20 Francs	0.187	1808-1811
	10 Francs	0.093	1813
	5 Francs	0.047	1813

Country or Coin Area		Gold Content in oz.	Years Produced
Germany	10 Gulden	0.199	1819-1842
(cont.)	5 Gulden	0.100	1819-1835
	1 Krone	0.321	1857-1871
	1/2 Krone	0.161	1857-1870
	15 Rupees	0.174	1916
	20 Marks	0.230	1871-1915
	10 Marks	0.115	1872-1914
	5 Marks	0.058	1877-1878
Ghana	5 Pounds	0.471	1960

Great Britain   *The weight of British coins prior to 1663 fluctuated greatly.
The gold contents and values quoted reflect averages.*

	1 Noble	0.251	1327-1483
	1 Angel	0.154	1422-1625
	1 Ryal	0.400	1485-1625
	1 Sovereign	0.384	1485-1525
	1 Sovereign	0.354	1526-1547
	1 Sovereign	0.384	1551-1625
	1 George Noble	0.133	1509-1547
	3 Pounds (Triple Unite)	0.796	1642-1644
	20 Shillings (One Unite)	0.265	1603-1663
	2 Crowns or 10 shillings	0.133	1603-1663
	1 Crown or 5 shillings	0.066	1509-1666
	1/2 Crown or 2 1/2 shillings	0.033	1509-1625
	5 Guineas	1.230	1668-1777
	2 Guineas	0.492	1664-1777
	1 Guinea	0.246	1663-1813
	1/2 Guinea	0.123	1669-1813
	1/3 Guinea	0.082	1797-1813
	1/4 Guinea	0.062	1718-1762
	5 Pounds	1.177	1820-1965
	2 Pounds	0.471	1820-1953
	1 Pound (Sov.)	0.235	1817-1966
	1/2 Pound (1/2 Sov.)	0.118	1817-1965
Greece	100 Drachmae	0.933	1876-1935
	50 Drachmae	0.467	1876

Country or Coin Area		Gold Content in oz.	Years Produced
Greece (cont.)	40 Drachmae	0.373	1852
	20 Drachmae	0.187	1833-1935
	10 Drachmae	0.093	1876
	5 Drachmae	0.047	1876
Guatemala	20 Pesos	0.933	1869-1878
	16 Pesos	0.747	1863-1869
	10 Pesos	0.467	1869
	8 Pesos	0.373	1864
	5 Pesos	0.233	1869-1878
	4 Pesos	0.187	1861-1869
	2 Pesos	0.093	1859
	1 Peso	0.047	1859-1860
	4 Reales ($^{1}/_{2}$ Peso)	0.023	1859-1864
	20 Quetzales	0.967	1926
	10 Quetzales	0.484	1926
	5 Quetzales	0.242	1926
Guinea, Equ.	5000 Pesetas	2.041	1970
	1000 Pesetas	.408	1970
	750 Pesetas	0.306	1970
	500 Pesetas	0.204	1970
	200 Pesetas	0.102	1970
Guinea	10000 Francs	1.157	1969
	5000 Francs	0.579	1969-1970
	2000 Francs	0.232	1969-1970
	1000 Francs	0.116	1969
Haiti	1000 Gourdes	5.715	1967-1969
	500 Gourdes	2.857	1968
	250 Gourdes	1.429	1968
	200 Gourdes	1.143	1967-1971
	100 Gourdes	0.572	1967-1971
	60 Gourdes	0.371	1968
	50 Gourdes	0.286	1967-1971
	40 Gourdes	0.248	1968
	30 Gourdes	0.186	1968
	20 Gourdes	0.114	1967-1969
	1000 Gourdes	0.421	1973
	500 Gourdes	0.211	1973
	200 Gourdes	0.084	1973
	100 Gourdes	0.042	1973
	1000 Gourdes	0.379	1974

Country or Coin Area		Gold Content in oz.	Years Produced
Haiti (cont.)	500 Gourdes	0.188	1974
	200 (Pope)	0.076	1975
	200 (Women)	0.085	1975
Hejaz	1 Dinar	0.213	1923
Honduras	20 Pesos	0.933	1888-1908
	10 Pesos	0.467	1871-1889
	5 Pesos	0.233	1871-1913
	1 Peso	0.047	1871-1922
	100 Lempiras	0.933	1971
	50 Lempiras	0.467	1971
	20 Lempiras	0.233	1971
	10 Lempiras	0.093	1971
Hong Kong	1000 Dollars		
	(Royal Visit)	0.471	1975
	(Dragon)	0.471	1976
	(Snake)	0.471	1977
	(Horse)	0.471	1978
	(Goat)	0.471	1979
	1 Ducat	0.111	1792-1831
	8 Forint	0.187	1870-1892
	4 Forint	0.093	1870-1892
Hungary	8 Florins	0.187	1870-1892
	4 Florins	0.093	1870-1892
	100 Korona	0.980	1907-1908
	20 Korona	0.196	1892-1918
	10 Korona	0.098	1892-1915
	100 Pengo	1.259	1938
	100 Pengo	0.839	1929-1938
	40 Pengo	0.336	1935
	20 Pengo	0.169	1928-1929
	10 Pengo	0.085	1928
	1000 Forint	2.433	1966
	500 Forint	1.111	1961
	500 Forint	1.217	1966
	200 Forint	0.487	1968
	100 Forint	0.222	1961
	100 Forint	0.243	1966
	50 Forint	0.111	1961
Iceland	10000 Kroner	0.449	1974

Country or Coin Area		Gold Content in oz.	Years Produced
Iceland (cont.)	5000 Kroner	0.259	1961
India	200 Mohurs	68.73	1628-1658
	100 Mohurs	34.365	1556-1707
	5 Mohurs	1.718	1556-1627
	2 Mohurs	0.687	1556-1835
	1 Mohur	0.344	1200-1947
	1/2 Mohur	0.172	1200-1947
	1/3 Mohur	0.115	1200-1947
	1/4 Mohur	0.086	1200-1947
	1/6 Mohur	0.057	1500-1820
	1/8 Mohur	0.043	1500-1820
	1/16 Mohur	0.022	1500-1820
	1/32 Mohur	0.011	1500-1820
	5 Rupees (1/3 Mohur)	0.115	1820-1879
	10 Rupees (2/3 Mohur)	0.230	1862-1879
	15 Rupees	0.235	1918
	1 Pagoda (primitive)	0.077	1200-1868
	2 Pagoda (modern) Madras	0.172	1810
	1 Pagoda (modern) Madras	0.086	1810
	2 Pagoda (modern) Travancore	0.150	1877-1924
	1 Pagoda (modern) Travancore	0.075	1877-1924
	1/2 Pagoda (modern) Travancore	0.038	1877-1924
	1/4 Pagoda (modern) Travancore	0.019	1877-1924
	100 Kori (Cuch Bhuj)	0.551	1866
	50 Kori (Cuch Bhuj)	0.276	1873-1874
	25 Kori (Cuch Bhuj)	0.138	1862-1870
Indonesia	25000 Rupiah	1.786	1970
	20000 Rupiah	1.429	1970

Country or Coin Area		Gold Content in oz.	Years Produced
Indonesia	10000 Rupiah	0.714	1970
(cont.)	5000 Rupiah	0.357	1970
	2000 Rupiah	0.143	1970
	100000 Rupiah	0.967	1974
Iraq	5 Dinars	0.400	1971
Isle of Man	5 Pounds	1.177	1965
	1 Sovereign	0.235	1965
	1/2 Sovereign	0.118	1965
	5 Pounds	1.173	1973-1974
	2 Pounds	0.469	1973-1974
	1 Sovereign	0.235	1973-1974
	1/2 Sovereign	0.117	1973-1974
Israel	20 Pounds	0.235	1960
	100 Pounds	0.736	1962-1967
	50 Pounds	0.393	1962-1964
	100 Pounds	0.643	1968-1969
	100 Pounds	0.637	1971
	200 Pounds	0.781	1973
	100 Pounds	0.391	1973
	50 Pounds	0.203	1973
	500 Pounds	0.810	1974
	500 Pounds	0.579	1975
Italy	1 Doppia (2 Ducats)	0.222	1200-1815
	1 Genovino (1 Ducat)	0.111	1200-1415
	1 Florin (1 Ducat)	0.111	1250-1500
	1 Scudo d'oro (1 Ducat)	0.111	1300-1750
	1 Zecchino (1 Ducat)	0.111	1500-1800
	1 Zecchino, Florence (= 1 Ducat)	0.112	1712-1853
	1 Ruspone, Florence (= 3 Zecchini)	0.336	1719-1859
	80 Florins, Florence	1.049	1827-1828

Country or Coin Area		Gold Content in oz.	Years Produced
Italy (cont.)	6 Ducats, Naples	0.248	1749-1785
	4 Ducats, Naples	0.165	1749-1782
	2 Ducats, Naples	0.827	1749-1772
	30 Ducats, Naples	1.213	1818-1856
	15 Ducats, Naples	0.606	1818-1856
	6 Ducats, Naples	0.243	1818-1856
	3 Ducats, Naples	0.121	1818-1856
Ivory Coast	100 Francs	0.926	1966
	50 Francs	0.463	1966
	25 Francs	0.232	1966
	10 Francs	0.926	1966
Jamaica	20 Dollars	0.253	1972
	100 Dollars	0.227	1975
Japan	20 Yen	0.965	1870-1880
	10 Yen	0.482	1871-1880
	5 Yen	0.241	1870-1897
	2 Yen	0.097	1870-1880
	1 Yen	0.048	1871-1880
	20 Yen	0.482	1897-1932
	10 Yen	0.241	1897-1910
	5 Yen	0.121	1897-1930
Jersey	50 Pounds	0.667	1972
	25 Pounds	0.351	1972
	20 Pounds	0.273	1972
	10 Pounds	0.137	1972
	5 Pounds	0.077	1972
Jordan	25 Dinars	2.000	1969
	10 Dinars	0.800	1969
	5 Dinars	0.400	1969
	2 Dinars	0.160	1969
Katanga	5 Francs	0.386	1961

Country or Coin Area		Gold Content in ozs.	Years Produced
Kenya	500 Shillings	1.120	1966
	250 Shillings	0.560	1966
	100 Shillings	0.224	1966
Korea	20 Won	0.482	1906-1910
	10 Won	0.241	1906-1910
	5 Won	0.121	1908-1909
Kuwait	5 Dinars	0.400	1961
Laos	80000 Kips	2.315	1971
	40000 Kips	1.157	1971
	20000 Kips	0.579	1971
	8000 Kips	0.232	1971
	4000 Kips	0.116	1971
	100000 Kips	0.219	1975
	50000 Kips	0.106	1975
Lesotho	20 Maloti	2.352	1966-1969
	10 Maloti	1.177	1966-1969
	4 Maloti	0.471	1966-1969
	2 Maloti	0.235	1966-1969
	1 Maloti	0.118	1966-1969
Liberia	20 Dollars (Red)	0.540	1964
	20 Dollars (Yellow)	0.599	1964
	30 Dollars	0.434	1965
	25 Dollars	0.675	1965
	12 Dollars	0.174	1965
	20 Dollars	0.968	1972
	10 Dollars	0.484	1972
	5 Dollars	0.242	1972
	2$1/2$ Dollars	0.121	1972
Liechtenstein	20 Kronen	0.196	1898-1900
	10 Kronen	0.097	1898-1900
	20 Franken	0.187	1930-1946
	10 Franken	0.093	1930-1946
	100 Franken	0.933	1952
	50 Franken	0.327	1956-1961
	25 Franken	0.163	1956-1961

Country or Coin Area		Gold Content in oz.	Years Produced
Luxembourg	20 Francs	0.187	1953
Malaysia	100 Ringgit	0.549	1971
Mali	100 Francs	0.926	1967
	50 Francs	0.463	1967
	25 Francs	0.232	1967
	10 Francs	0.093	1967
Malta	20 Scudi	0.432	1764-1778
	10 Scudi	0.216	1756-1782
	5 Scudi	0.108	1756-1779
	50 Pounds	0.965	1972
	20 Pounds	0.358	1972
	10 Pounds	0.177	1972
	5 Pounds	0.088	1972
	50 Pounds	0.442	1973-1975
	20 Pounds	0.177	1973-1975
	10 Pounds	0.088	1973-1975
Mauritius	200 Rupees	0.459	1971
	1000 Rupees	0.471	1978
Mexico	8 Escudos	0.762	1822
	8 Escudos	0.762	1823
	8 Escudos (Eagle)	0.762	1823
	8 Escudos (facing Eagle)	0.762	1824-1873
	4 Escudos	0.381	1823
	4 Escudos	0.381	1825-1869
	2 Escudos	0.190	1825-1870
	1 Escudo	0.095	1825-1870
	$1/2$ Escudo	0.048	1825-1869
	20 Pesos	0.952	1870-1905
	10 Pesos	0.476	1870-1905
	5 Pesos	0.238	1870-1905
	$2^{1}/2$ Pesos	0.119	1870-1893
	1 Peso	0.048	1870-1905
	50 Pesos	1.206	1921-1947
	20 Pesos	0.482	1917-1959
	10 Pesos	0.241	1905-1959
	5 Pesos	0.121	1905-1955

Country or Coin Area		Gold Content in oz.	Years Produced
Mexico	2¹/₂ Pesos	0.060	1918-1948
(cont.)	2 Pesos	0.048	1919-1948
Monaco	40 Francs	0.373	1838
	100 Francs	0.933	1882-1904
	20 Francs	0.187	1838-1892
	2 Francs	0.463	1943
	1 Franc	0.231	1943
	100 Francs	0.738	1950
	100 Francs (Double)	1.476	1950
	50 Francs	0.593	1950
	50 Francs (Double)	1.186	1950
	20 Francs (Double)	0.420	1950
	20 Francs (Double)	0.839	1950
	10 Francs	0.304	1950
	10 Francs (Double)	0.608	1950
	100 Francs	0.347	1956
	200 Francs	0.947	1966
Montenegro	100 Perpera	0.980	1910
	20 Perpera	0.196	1910
	10 Perpera	0.098	1910
Morocco	4 Ryals	0.187	1879
	250 Dirham	0.187	1977
Muscat and Oman	15 Ryals	0.235	1962
Nepal	4 Mohars	0.679	1750-1938
	2 Mohars	0.340	1750-1938
	1 Mohar	0.170	1750-1938
	¹/₂ Mohar	0.085	1750-1938
	¹/₄ Mohar	0.042	1750-1911
	¹/₈ Mohar	0.021	1750-1911
	¹/₁₆ Mohar	0.011	1750-1911
	¹/₃₂ Mohar	0.005	1750-1911
	¹/₆₄ Mohar	0.003	1750-1911
	¹/₁₂₈ Mohar	0.001	1881-1911

Country or Coin Area		Gold Content in oz.	Years Produced
Nepal (cont.)	2 Rupees	0.340	1948-1955
	1 Rupee	0.170	1938-1962
	1/2 Rupee	0.085	1938-1962
	1/4 Rupee	0.086	1955
	1/5 Rupee	0.069	1955
	1/6 Rupee	0.062	1956
Netherlands	14 Guilders	0.293	1749-1764
	7 Guilders	0.146	1749-1764
	20 Guilders	0.402	1808-1810
	10 Guilders	0.201	1808-1810
	20 Guilders	0.397	1848-1853
	10 Guilders	0.195	1818-1933
	5 Guilders	0.097	1826-1912
	1 Ducat	0.110	1960-1972
Netherlands East Indies	1 Ducat	0.110	1814-1937
Newfoundland	2 Dollars	0.098	1865-1888
Nicaragua	2000 Cordobas	0.500	1975
	1000 Cordobas	0.278	1975
	500 Cordobas	0.156	1975
	200 Cordobas	0.061	1975
Niger	100 Francs	0.926	1960-1968
	50 Francs	0.463	1960-1968
	25 Francs	0.232	1960-1968
	10 Francs	0.093	1960-1968
Norway	20 Kronor	0.259	1874-1910
	10 Kronor	0.130	1874-1910
Oman	500 Ryals	1.179	1971
	200 Ryals	0.472	1971
	100 Ryals	0.236	1971
	50 Ryals	0.118	1971
Ottoman Empire	5 Sequins	0.386	1451-1839
	4 Sequins	0.389	1451-1839
	3 Sequins	0.231	1451-1839
	2 Sequins	0.154	1451-1839
	1 Sequin	0.077	1451-1839
	1/2 Sequin	0.039	1451-1839

Country or Coin Area		Gold Content in oz.	Years Produced
Panama	100 Balboas	0.236	1974
	500 Balboas	1.207	1975
Paraguay	10000 Guaranies	1.331	1968
	4500 Guaranies	0.923	1972-1974
	3000 Guaranies	0.616	1972-1974
	1500 Guaranies	0.310	1972-1974
Persia	1 Ashrafi	0.110	1500-1750
	25 Tomans	2.081	1848-1896
	20 Tomans	1.664	1848-1986
	20 Tomans	0.832	1848-1925
	5 Tomans	0.418	1848-1925
	2 Tomans	0.166	1848-1925
	1 Toman	0.083	1848-1927
	$^1/_2$ Toman	0.042	1848-1925
	$^1/_5$ Toman	0.017	1848-1925
	5 Pahlevi	0.278	1927-1930
	2 Pahlevi	0.111	1927-1930
	1 Pahlevi	0.056	1927-1930
	5 Pahlevi	1.177	1961-1975
	$2^1/_2$ Pahlevi	0.589	1961-1975
	1 Pahlevi	0.235	1932-1975
	$^1/_2$ Pahlevi	0.118	1932-1975
	$^1/_4$ Pahlevi	0.059	1950-1975
	2,000 Rials	0.754	1971
	1,000 Rials	0.377	1971
	750 Rials	0.283	1971
	500 Rials	0.188	1971
Peru	20 Soles	0.933	1863
	10 Soles	0.467	1863
	5 Soles	0.233	1863
	1 Libra	0.235	1898-1964
	$^1/_2$ Libra	0.118	1902-1964
	$^1/_5$ Libra	0.047	1906-1964
	50 Soles	0.968	1930-1931
	100 Soles	1.354	1950-1969
	50 Soles	0.677	1950-1969
	20 Soles	0.271	1950-1969
	10 Soles	0.135	1956-1969
	5 Soles	0.68	1956-1969
Philippines	4 Pesos	0.190	1861-1882

Country or Coin Area		Gold Content in oz.	Years Produced
Philippines	2 Pesos	0.095	1861-1868
(cont.)	1 Peso	0.048	1861-1868
	1 Piso	0.570	1970
	1000 Pisos	0.288	1975
Poland	1 Ducat	0.111	1812-1831
	50 Zloty	0.289	1817-1829
	25 Zloty	0.144	1817-1833
	20 Zloty (3 Roubles)	0.109	1834-1840
	20 Zloty	0.187	1925
	10 Zloty	0.093	1925
Portugal	4 Cruzados	0.444	1580-1652
	2 Cruzados	0.222	1580-1647
	1 Cruzado	0.111	1438-1647
	4000 Reis	0.316	1663-1722
	2000 Reis	0.158	1663-1725
	1000 Reis	0.079	1663-1821
	400 Reis	0.032	1717-1821
	8 Escudos	0.843	1717-1732
	4 Escudos	0.422	1722-1835
	2 Escudos	0.211	1722-1831
	1 Escudo	0.105	1722-1821
	$^1/_2$ Escudo	0.053	1722-1821
	5000 Escudos	0.282	1836-1851
	2500 Escudos	0.141	1838-1853
	1000 Escudos	0.070	1851
	10000 Escudos	0.523	1878-1889
	5000 Escudos	0.261	1860-1889
	2000 Escudos	0.105	1856-1888
	1000 Escudos	0.052	1855-1879
Ras Al	200 Riyals	1.198	1970
Khaima	150 Riyals	0.899	1970
	100 Riyals	0.599	1970
	75 Riyals	0.449	1970
	50 Riyals	0.300	1970
Rhodesia	5 Pounds	1.177	1966
	1 Pound	0.235	1966
	10 Shillings	0.118	1966
Romania	100 Lei	0.933	1906-1940

Country or Coin Area		Gold Content in oz.	Years Produced
Romania	50 Lei	0.467	1906-1922
(cont.)	25 Lei	0.233	1906-1922
	20 Lei	0.187	1867-1944
	12^1/$_2$ Lei	0117	1906
Russia	25 Roubles	0.964	1876
	10 Roubles	0.384	1836
	5 Roubles	0.193	1826-1885
	3 Roubles	0.115	1826-1885
	10 Roubles	0.373	1886-1894
	5 Roubles	0.187	1886-1894
	15 Roubles	0.373	1897
	7^1/$_2$ Roubles	0.187	1897
	37^1/$_2$ Roubles	0.933	1902
	25 Roubles	0.933	1896-1908
	10 Roubles	0.249	1898-1923
	5 Roubles	0.125	1897-1910
	100 Roubles (Olympic)	0.500	1979-1980
Rwanda	100 Francs	0.868	1965
	50 Francs	0.434	1965
	25 Francs	0.217	1965
	10 Francs	0.087	1965
Salvador	20 Pesos	0.933	1892
	10 Pesos	0.467	1892
	5 Pesos	0.233	1892
	2^1/$_2$ Pesos	0.117	1892
	20 Colones	0.450	1925
	200 Colones	0.683	1971
	100 Colones	0.341	1971
	50Colones	0.171	1971
	25 Colones	0.085	1971
San Marino	20 Lire	0.187	1925
	10 Lire	0.093	1925
	2 Scudi	0.177	1974
	1 Scudo	0.088	1974
Saudi Arabia	4 Saudi Pounds	0.942	1945-1946
	1 Saudi Pound	0.235	1951-1957
Senegal	100 Francs	0.926	1968

Country or Coin Area		Gold Content in oz.	Years Produced
Senegal	50 Francs	0.463	1968
(cont.)	25 Francs	0.232	1968
	10 Francs	0.093	1968
Serbia	20 Dinars	0.187	1879-1882
	10 Dinars	0.093	1882
Sharjah	200 Riyals	1.200	1970
	100 Riyals	0.600	1970
	50 Riyals	0.300	1970
	25 Riyals	0.150	1970
Siam	8 Ticals	0.227	1851-1868
	4 Ticals	0.113	1851-1868
	2 Ticals	0.057	1851-1907
Sierra	1 Golde	1.578	1966
Leone	1/2 Golde	0.789	1966
	1/4 Golde	0.395	1966
Singapore	150 Dollars	0.736	1969
	500 Dollars	1.000	1975
	250 Dollars	0.500	1975
	100 Dollars	0.200	1975
Somalia	500 Shillings	2.026	1965-1970
	200 Shillings	1.013	1965-1970
	100 Shillings	0.506	1965-1970
	50 Shillings	0.253	1965-1970
	20 Shillings	0.127	1965-1970
South	1 Pound	0.235	1874-1902
Africa	1/2 Pound	0.118	1892-1897
	1 British Sovereign	0.235	1923-1960
	1/2 British Sovereign	0.118	1923-1960
	2 Rand	0.235	1961-1964
	1 Rand	0.118	1961-1964
	1 Krugerrand	1.000	since 1967
	1/2 Krugerrand	0.500	since 1980
	1/4 Krugerrand	0.250	since 1980
	1/10 Krugerrand	0.100	since 1980
South Korea	25000 Won	2.801	1970
	20000 Won	2.240	1970

Country or Coin Area		Gold Content in oz.	Years Produced
South Korea	10000 Won	1.120	1970
(cont.)	5000 Won	0.560	1970
	2000 Won	0.280	1970
	1000 Won	0.112	1970
Spain	10 Doblas	1.106	1350-1369
	5 Doblas	0.713	1454-1474
	1 Dobla	0.143	1252-1474
	¹/2 Dobla	0.071	1454-1474
	1 Excelente	0.111	1476-1516
	8 Escudos	0.796	1516-1772
	4 Escudos	0.398	1516-1772
	2 Escudos	0.199	1516-1772
	1 Escudo	0.099	1516-1772
	8 Escudos	0.781	1773-1785
	4 Escudos	0.391	1773-1785
	2 Escudos	0.195	1773-1785
	1 Escudo	0.098	1773-1785
	8 Escudos	0.076	1786-1833
	4 Escudos	0.380	1786-1833
	2 Escudos	0.190	1786-1833
	1 Escudo	0.095	1786-1833
	320 Reales	0.760	1810-1823
	160 Reales	0.380	1822
	80 Reales	0.190	1809-1848
	100 Reales	0.242	1850-1868
	40 Reales	0.097	1861-1868
	20 Reales	0.048	1861-1865
	100 Pesetas	0.933	1870-1897
	25 Pesetas	0.233	1871-1885
	20 Pesetas	0.187	1889-1904
	10 Pesetas	0.093	1878
Spanish	8 Escudos	0.760	1598-1873
America	4 Escudos	0.380	1598-1873
	2 Escudos	0.190	1598-1873
	1 Escudo	0.095	1598-1873
	¹/2 Escudo	0.066	1598-1873
Swaziland	1 Lilangeni	0.836	1968
	25 Emalangeni	0.804	1974
	20 Emalangeni	0.643	1974
	10 Emalangeni	0.322	1974
	5 Emalangeni	0.161	1974
Sweden	1 Ducat	0.111	1793-1868

Country or Coin Area		Gold Content in oz.	Years Produced
Sweden (cont.)	2 Ducats	0.221	1836-1857
	1 Carolin (10 francs)	0.093	1868-1872
	20 Kronor	0.259	1873-1925
	10 Kronor	0.130	1873-1901
	5 Kronor	0.065	1881-1920
Switzerland	24 Münzgulden	0.442	1794-1796
	12 Münzgulden	0.221	1794-1796
	6 Duplonen	1.326	1794
	4 Duplonen	0.884	1797-1798
	2 Duplonen	0.442	1793-1798
	1 Duplonen	0.221	1787-1829
	$1/2$ Duplonen	0.111	1787-1796
	$1/4$ Duplonen	0.006	1789-1796
	32 Franken	0.442	1800
	16 Franken	0.221	1800-1813
	8 Franken	0.111	1813
	20 Francs (Geneva)	0.183	1848
	10 Francs (Geneva)	0.092	1848
	10 Francs	0.093	1911–1922
	20 Francs	0.187	1871–1947
	100 Francs	0.933	1925
	100 Francs	0.749	1934
	100 Francs	0.506	1939
Syria	1 Pound	0.195	1950
	$1/2$ Pound	0.098	1950
Tanzania	1500 Shillings	0.968	1974
Thailand	600 Baht	0.434	1968
	300 Baht	0.217	1968
	150 Baht	0.109	1968
	800 Baht	0.579	1971
	400 Baht	0.289	1971
Tonga	1 Koula	0.959	1962
	$1/2$ Koula	0.479	1962
	$1/2$ Koula	0.240	1961
Tunis	100 Piastres	0.564	1855-1864
	80 Piastres	0.451	1855
	50 Piastres	0.282	1855-1867

Country or Coin Area		Gold Content in oz.	Years Produced
Tunis (cont.)	40 Piastres	0.226	1855
	20 Piastres	0.141	1857-1882
	10 Piastres	0.113	1855
	5 Piastres	0.056	1855-1871
	2¹/₂ Piastres	0.028	1864-1872
	20 Francs	0.187	1891-1928
	15 Francs	0.140	1886-1891
	10 Francs	0.093	1891-1928
	100 Francs	0.190	1930-1955
Tunisia	40 Dinars	2.200	1967
	20 Dinars	1.100	1967
	10 Dinars	0.550	1967
	5 Dinars	0.275	1967
	2 Dinars	0.110	1967
Turkey	5 Sequins	0.386	1703-1839
	4 Sequins	0.309	1703-1839
	3 Sequins	0.232	1703-1839
	2 Sequins	0.154	1703-1839
	1 Sequin	0.077	1451-1839
	¹/₂ Sequin	0.039	1451-1839
	¹/₄ Sequin	0.019	1451-1839
	500 Piastres	1.063	1839-1975
	250 Piastres	0.532	1839-1975
	100 Piastres	0.213	1839-1975
	50 Piastres	0.106	1839-1975
	25 Piastres	0.053	1839-1975
	12¹/₂ Piastres	0.027	1909-1918
	500 Piastres (deluxe)	1.034	SINCE 1926
	250 Piastres	0.517	SINCE 1926
	100 Piastres	0.207	SINCE 1926
	50 Piastres	0.103	SINCE 1926
	25 Piastres	0.052	SINCE 1926
	500 Lire	0.177	1973
Turks & Caicos Islands	100 Crowns	0.290	1974
	50 Crowns	0.145	1974
	100 Crowns	0.200	1975
	50 Crowns	0.100	1975
	25 Crowns	0.050	1975
Tuvala	50 Dollars	0.471	1976

Country or Coin Area		Gold Content in oz.	Years Produced
Uganda	1000 Shillings	4.000	1969
	500 Shillings	2.000	1969
	100 Shillings	0.400	1969
	50 Shillings	0.200	1969
Um-Al-Qawain	200 Riyals	1.200	1970
	100 Riyals	0.600	1970
	50 Riyals	0.300	1970
	25 Riyals	0.150	1970
United States of America	10 Dollars	0.516	1795-1804
	5 Dollars	0.258	1795-1833
	2^1/$_2$ Dollars	0.129	1796-1833
	5 Dollars	0.242	1834-1836
	2^1/$_2$ Dollars	0.121	1834-1836
	10 Dollars	0.483	1838-1933
	5 Dollars	0.242	1837-1929
	2^1/$_2$ Dollars	0.121	1837-1929
	20 Dollars	0.968	1850-1933
	3 Dollars	0.145	1854-1889
	1 Dollar	0.048	1849-1889
	4 Dollars	0.194	1879-1880
	50 Dollars	2.419	1915
Uruguay	5 Pesos	0.250	1930
Vatican	4 Doppia	0.631	1786-1787
	2 Doppia	0.315	1776-1777
	1 Doppia	0.158	1776-1834
	^{1}/$_2$ Doppia	0.079	1776-1787
	10 Scudi	0.501	1835-1856
	5 Scudi	0.250	1835-1854
	2^1/$_2$ Scudi	0.125	1835-1863
	1 Scudo	0.050	1853-1865
	100 Lire	0.933	1866-1870
	50 Lire	0.467	1868-1870
	20 Lire	0.187	1866-1870
	10 Lire	0.093	1866-1869
	5 Lire	0.047	1866-1867
	100 Lire	0.255	1929-1935
	100 Lire	0.150	1936-1959
Venezuela	100 Bolivares	0.933	1875-1889
	50 Bolivares	0.467	1875-1888

Country or Coin Area		Gold Content in oz.	Years Produced
Venezuela	25 Bolivares	0.233	1875
(cont.)	20 Bolivares	0.187	1879-1912
	10 Bolivares	0.093	1930
	5 Bolivares	0.047	1875
Yemen Arab	50 Ryals	1.418	1969
Republic	30 Ryals	0.851	1969
	20 Ryals	0.567	1969
	10 Ryals	0.284	1969
	5 Ryals	0.142	1969
Yugoslavia	20 Dinars	0.187	1925
	1000 Dinars	2.263	1968
	500 Dinars	1.131	1968
	200 Dinars	0.453	1968
	100 Dinars	0.226	1968
Zanzibar	5 Rials	0.242	1881
	$2^1/_2$ Rials	0.121	1881

Bibliography

There is a vast body of technical, semi-technical, historical and contemporary literature on the different aspects of the study of gold. In order to cater thoroughly for all interests it would be necessary, in compiling even a select list of books to record many hundreds of titles. The list which follows attempts no such comprehensivity and is in no sense exhaustive. It is divided into five main categories: general interest books, the role of gold in the world's monetary system and specialist topics. Many of the books do themselves have extensive bibliographies.

General Books

Blakemore, K. *The Book of Gold*, New York: Stein & Day, 1971.

Busschau, W.J. *Measure of Gold*, Central News Agency, Johannesburg, 1949.

Cartwright, A.P. *The Gold Miners*, Purnell & Sons, Cape Town, 1962.

Cartwright, A.P. *Gold Paved the Way: The Story of the Gold Fields Group of Companies*, Macmillan, London, St. Martin's Press, New York, 1967.

Cartwright, A.P. *West Driefontein—Ordeal by Water*, Gold Fields of South Africa, 1969.

Einzig, P. *Primitive Money*, Eyre and Spottiswoode, London, 1948.

Green, T.S. *The World of Gold Today*, New York, Walker, 1973.

Holmyard, E.J. *Alchemy*, Penguin Books, Harmondsworth, 1957.

Robbins, P. and Lee, D. *Guide to Precious Metals and Their Markets*. Kogan Page, London, 1979.

Sutherland, C.H.V. *Gold: Its Beauty, Power and Allure*, Thames and Hudson, London, 1959.

Wise, Edmund M. *Gold: Recovery, Properties, and Applications*, Van Ostrand, Princeton, N.J., 1964.

Wilson, F. *Labour in the South African Gold Mines 1911–1969*, Cambridge University Press, 1972.

The History of Gold (I)

Aitchison, L. *A History of Metals*, 2 vols., Macdonald and Evans, London, 1960.

Allen, G. *Gold, History from Ancient Times to the Present Day*, New York, 1965.

Anderson, R.S. *Australian Goldfields*, D.S. Ford, Sydney, 1956.

Berton, P. *The Golden Trail*, Macmillan, Toronto, 1954.

Bolin, S. *State and Currency in the Roman Empire*, Almquist and Wiksell, Stockholm, 1958.

The life of Benvenuto Cellini, written by himself, Phaidon Press Ltd., London, 1949.

Clapham, Sir John, *The Bank of England*, 2 vols., University Press, Cambridge, 1944.

Gray, J. and E. *A History of the Discovery of the Witwatersrand Gold Fields*, Sholto Douglas, 1940.

Healey, J.F. *Mining and Metallurgy in the Greek and Roman World*, Thames and Hudson, 1978.

Hyams, E. and Ordish, G. *The Last of the Incas*, Longmans Green, London, 1963.

Hemming, J. *The Search for El Dorado*, Michael Joseph, 1978.

Jastram, R.W. *The Golden Constant: The English and American Experience, 1560-1976*, Wiley, New York, 1977.

Morell, W.P. *The Gold Rushes*, A. & C. Black, London, 1940.

Murray, A.E. *Murray's Guide to the Gold Diggings*, D.S. Ford, Sydney, 1956.

Paul, R.W. *Californian Gold*, Harvard University Press, Cambridge, Mass., 1947.

Richards, P.D. *The Early History of Banking in England.*

Sédillot, R. *Histoire de l'Or*, Librairie Arthene Fayard, France, 1972.

Shinn, C.H. *The Story of the Mine: as illustrated by the Great Comstock Lode of Nevada*, University of Nevada Press, 1980.

The History of the Comstock Lode 1850-1920, Mackay School of Mines, University of Nevada.

The Travels of Marco Polo, Modern Library, New York, 1953.

Vilar, P. *A History of Gold and Money (1450-1920).*

The History of Gold (II): The Gold Standard Periods

Bergsten, C.F. *The Dilemmas of the Dollar*, New York University Press, 1975.

Bloomfield, A.I. *Short-Term Capital movements under the 1914 Gold Standard*, Princeton Papers No. 11.

Brown, W.A. Jr. *England and the New Gold Standard, 1919-1926*, New Haven, 1929.

Brown, W.A. Jr. *The International Gold Standard Reinterpreted, 1914-1934*, N.Y., 1940.

Cassel, G. *The Downfall of the Gold Standard*, Clarendon Press, Oxford, 1936.

Clarke, S.O.V. *Central Bank Co-operation, 1924-31*, N.Y., 1967.

Committee on Finance and Industry (Macmillan Committee), *Report*, Cmd. 1897, *Evidence* (1931).

Einzig, P. *The Tragedy of the Pound*, 1932.

Gregory, T.E. *The First Year of the Gold Standard*, 1920.

Harris, S. *John Maynard Keynes*, 1955.

Harrod, R.F. *The Life of John Maynard Keynes*, 1951.

Hawtrey, R.G. *The Gold Standard in Theory and Practice*, 1927, 4th ed. 1939.

Jones, J.H. 'The Gold Standard', *Econ. J.* 43/172, December, 1933.

Keynes, J.M. *Essays in Persuasion*, 1931.

McKenna, R. *Post-War Banking Policy. A Series of Addresses*, 1928.

Moggridge, D.E. *The Return to Gold, 1925. The Formation of Economic Policy and its Critics*, Cambridge, 1969.

Moreau, E. *Souvenirs d'un Gouverneur de la Banque de France*, Paris, 1954.

Morgan Webb, Sir Charles, *Ten Years of Currency Revolution*, 1935.

Morgan, E.V. *Studies in British Financial Policy, 1914-25*, 1952.

Nevin, E. *The Mechanism of Cheap Money: a Study of British Monetary Policy, 1931-39*, Cardiff, 1955.

Pigou, A.C. *Aspects of British Economic History, 1918-1925*, 1947.

Sayers, R.S. 'The Return to Gold, 1925' in L.S. Presswell (ed) *Studies in the Industrial Revolution*, 1960.

Yaeger, L.B. *International Monetary Relations: Theory, History and Policy*, Harper and Row, 1976.

The Monetary Role of Gold

Bergsten, C.F. *The Dilemmas of the Dollar*, New York University Press, 1975.

Birnbaum, E.A. *Gold and the International Monetary System: An Orderly Reform*, Princeton Papers No. 66, April 1968.

Busschau, W.J. *Gold and International Liquidity*, South African Institute of International Affairs, Johannesburg, 1971.

Cassell, F. *Gold or Credit*, Pall Mall Press, London, 1965.

Consolidated Gold Fields Annual Report of Gold Market Developments.

Einzig, P. *The Destiny of Gold*, Macmillan, London, 1972.

Garritsen de Vries, M., The International Monetary Fund 1966-71, The System Under Stress, Vol. 1, Narrative Washington DC IMF, 1976.

Gilbert, M. *The Gold-Dollar System: Conditions of Equilibrium and the Price of Gold*, Princeton Papers No. 70, October 1968.

Gold and World Monetary Problems, Proceedings of the National Industrial Conference Board Convocation, Tarrytown, New York, October, 1965, published by Macmillan, New York, Collier-Macmillan, London, 1966.

Hayek, F.A. *Choice in Currency: A Way to Stop Inflation*, Levittown, N.Y., Transatlantic, 1977.

Hayek, F.A. *Denationalization of Money*, Institute of Economic Affairs, London, 1976.

Hirsch, F. *Money International*, Pelican Books, 1969.

International Monetary Fund: International Financial Statistics and the Annual Report.

Kriz, M.A. *Gold: Barbarous Relic or Useful Instrument?* Princeton Papers No. 60, June 1967.

Machlup, F. *The Book Value of Monetary Gold*, Princeton Papers No. 91, December 1971.

Morgan, E.V. *Gold Or Paper?* An essay on governments attempts to manage the post-war monetary system, and the case for and against restoring a link with gold, Institute of Economic Affairs, London, Hobart Paper 69, 1976.

Meiselman, D.I. and Laffer, A.B., eds. *The Phenomenon of Worldwide Inflation*, American Enterprise Institute, Washington, 1975.

Rees-Mogg, W. *The Reigning Error: the Crisis of World Inflation*, Hamilton, London, 1974.

Samuel Montagu Annual Bullion Review.

Rueff, J. and Hirsch, F. *The Role and the Rule of Gold: An Argument.* Princeton Papers No. 47, June 1965.

Triffin, R. *Gold and the Dollar Crisis*, Yale University Press, New Haven, Conn., and London, 1971.

Williamson, J. *The Failure of World Monetary Reform 1971–1974*, Thomas Nelson and Sons, 1977.

Young, J.P. *United States Gold Policy: The Case for Change*, Princeton Papers No. 56, October 1966.

Specialist Topics

Bayer, E. 'Gold from the Oceans' in *Chemistry* Vol. 37, No. 10, October 1964.

Carter, H. *The Tomb of Tut·ankh·Amen*, London, 1923–33.

Cavalti, P.C. *How to invest in Gold*, A Maximus Press Book, 1979.

Emmons, W.H. *Gold Deposits of the World: With a Section on Prospecting*, Arno, 1974. (Reprint of McGraw-Hill edition, 1937).

Fregnac, C. *Jewelry: From the Renaissance to Art Nouveau*, Octopus

Books, Ltd., London, 1973.

Friedberg, R. *Gold Coins of the World*, The Coin and Currency Institute, New York, 3rd Ed., 1971.

Grieveson, Grant and Co. (Stockbrokers), *Mining Quarterly*.

Jarecki, H.G. *The Success of the U.S. Gold Markets*, a paper presented to the *Financial Times* Gold Conference in Montreux in June 1979.

Kenyon, R.L. *Gold Coins of England*, Bernard Quaritch, 1884, Firecrest, 1970.

Laffer, A.B., Associates, *The Monetary Crisis: A Classical Perspective*.

Pick, F. *Gold—How and Where to Buy and Hold It*, Pick Publishing, New York, 1973.

Rapson and Groenewald, *Gold Usage*, 1978.

Rist, C. *The Triumph of Gold*, Greenwood, New York, 1969.

Rosenthal, E. *Gold! Gold! Gold!* Macmillan, New York, 1970.

Sarnoff, P. *Trading in Gold*, Woodhead-Faulkner, 1980.

Schlumberger, H. *Gold Coins of Europe Since 1800*, Sterling Publishing Co., New York, 1968.

Turney, J.C. *Economic Study: Gold*, published by A.B. Laffer Associates.

Index

Z